THE CAMCORDER

CHRIS GEORGE

Hodder & Stoughton

A MEMBER OF THE HODDER HEADLINE GROUP

British Library Cataloguing in Publication Data
A catalogue record for this title is available from the British Library

ISBN 0 340 59750 X

First published 1994
Impression number   11 10 9 8 7 6 5 4 3 2
Year   1999   1998   1997   1996   1995

Typeset by Rowland Phototypesetting Ltd, Bury St Edmunds, Suffolk.
Printed in Great Britain for Hodder & Stoughton Educational, a division of Hodder Headline Plc., 338 Euston Road, London NW1 3BH by Cox & Wyman Ltd, Reading, Berkshire.

Contents

—— INTRODUCTION ——

What is a camcorder?

There is something magical about seeing yourself on the television. Even the most introvert member of the family will be enthralled to see themselves captured by the camcorder.

The truth is that even the simplest camcorder can record images that mimic life so accurately that photographs seem positively prehistoric. A good video brings memories alive – showing every mannerism that makes our loved ones so special.

Not surprisingly, the camcorder has enjoyed a phenomenal rise in popularity in the last few years. The first camcorder, the Sony Betamovie, was launched in 1983. By 1993, 500,000 of these instant movie-making machines were being sold a year in the UK, and nearly three million a year in the US.

But what is a camcorder? Basically it is what it says – a CAMera and a video tape reCORDER rolled into one. The name comes about because in the early days of video, in the late 1970s, the recorder and the camera were two separate units. A heavy tape deck was carted around on a shoulder strap – with a wire stretching to a cumbersome video camera.

Today all domestic video cameras are camcorders – and many are small enough to slip into your coat pocket.

Recording moving images is not new. A generation ago our parents used cine cameras to record us growing up – but there were disadvantages. There was no sound. Film was expensive and lasted only long enough to give a few minutes footage. It had to be sent off for processing; a week or two later, when the film returned, a projector had

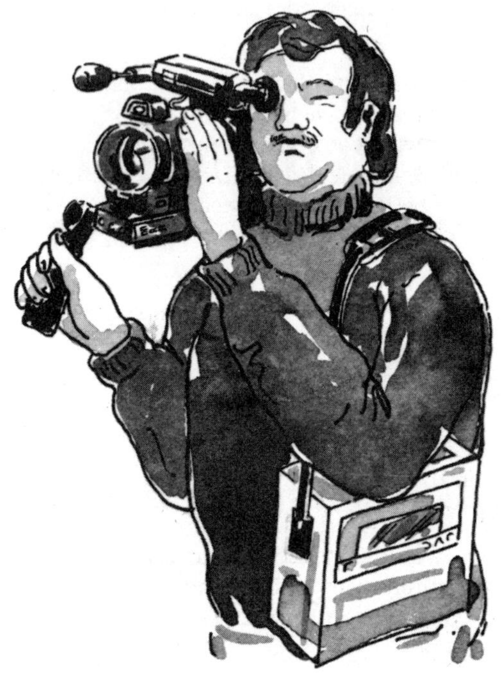

Early videomaking involved carrying a heavy video recorder on your shoulder and a separate camera unit

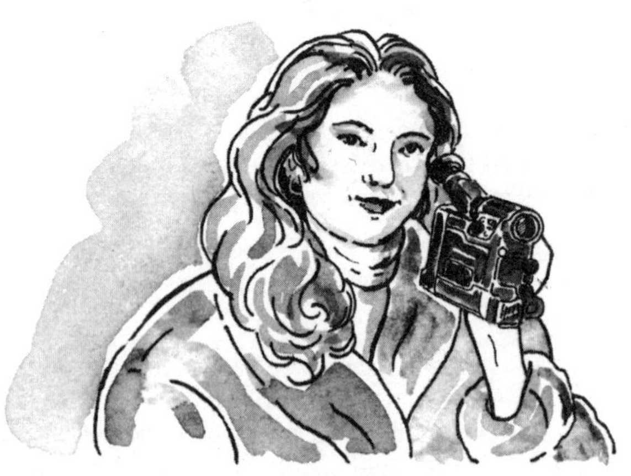

Some modern-day camcorders are so small and light that they can fit into a coat pocket

to be set up in a darkened room so that you could see the fruits of your labour.

The camcorder, on the other hand, records sound and full colour pictures that are perfectly synchronised – and that can be instantly replayed in the viewfinder of the camcorder, or, via a simple lead, straight on your TV screen. What's more, camcorders, once bought, are cheap to run. The batteries are rechargeable and the tapes can not only record several hours of footage, but they can also be re-recorded over again and again.

It would be true to say that camcorders are remarkably easy to use – and that even the cheapest is capable of staggering results. So why the need for this book? The answer is to avoid disappointment in the long run. All camcorders record moving pictures and sound – but the difference between models is immense. A camcorder is still an expensive purchase and not one that you want to replace in a year or two's time simply because you have grown beyond it. Secondly, there is a world of difference between using a camcorder and making a video that other people will want to watch. People are conditioned by the standards set by broadcast television – and will soon tire of amateurish productions.

This book, therefore, sets out to do two things. First, it describes the difference between the many models that you can buy. This is designed not only to help those who have yet to buy a camcorder to form an invaluable shopping list of useful features to look for, it is also to help the camcorder owner to become fully familiar with what all those buttons are used for – and how to get the best use out of them.

Secondly, the book describes many of the techniques that can be used to make your videos look, and sound, interesting. Video has its own language – and this doesn't come automatically to most people. It has to be learnt. From showing you how to frame your subjects in the viewfinder, through to how to edit your videos to keep your viewers awake and glued to their seats, this book will teach you all you need to know.

Happy reading – and happy shooting!

1
TYPES OF CAMCORDER

Camcorders come in all shapes and sizes, and whether you should go for a big professional-looking machine or the smallest one that you can find is largely a matter of personal taste. There is no right or wrong – the key is to find a model with which you feel comfortable.

Essentially there are three common types of camcorder: the on-the-shoulder type, the shoe-shaped and the palm-sized. Each has its advantages.

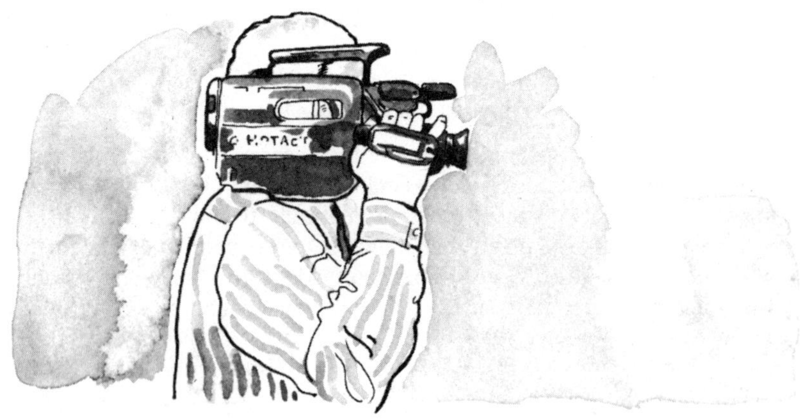

An on-the-shoulder camcorder

The on-the-shoulder type is the one that most closely resembles the ENG (electronic news gathering) cameras used by professional television crews. Despite the fact that it is the biggest and the heaviest type of camcorder available, it is not necessarily the least comfortable to hold. Although it can weigh up to 4kg, most of the weight is supported on the shoulder – so that your arms are relatively free from strain. Because of this, the camcorder is kept very stable – and your recordings are fairly immune from involuntary body movements. There is little worse in a movie than the effects caused by a wobbly camera operator. The most expensive camcorders tend to be found in this category.

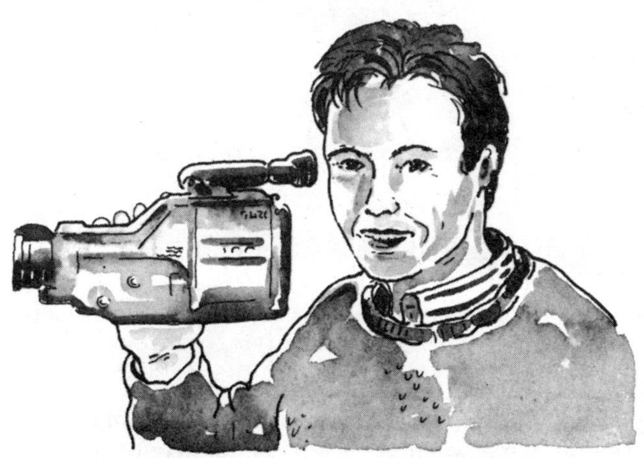

A typical shoe-shaped camcorder

The next size down is the shoe-shaped camcorder – so-called because from above they resemble the outline of a shoe. These are roughly twice as long as they are high or broad. The main weight of these is normally supported via an adjustable handgrip on the right-hand side of the camera. The left hand is then used to steady the camera from the side or underneath. Weighing 1–1.5kg they are significantly lighter than the on-the-shoulder type – but you must not forget that all the weight is being supported by the arms. Holding your hands in front of your face for more than a few minutes is more difficult than you may think.

The smallest type of camcorder is the palm-sized model or 'Palmcorder'. These can weigh between 600g and 1kg – and can be small enough to fit into a coat pocket or large handbag. Because they are so light, they can seemingly be held with one hand. However, because they are lightweight, they are also tend to be susceptible to camera shake – and therefore the left hand should still be used to steady the camcorder. Not surprisingly, the convenience of the small size of

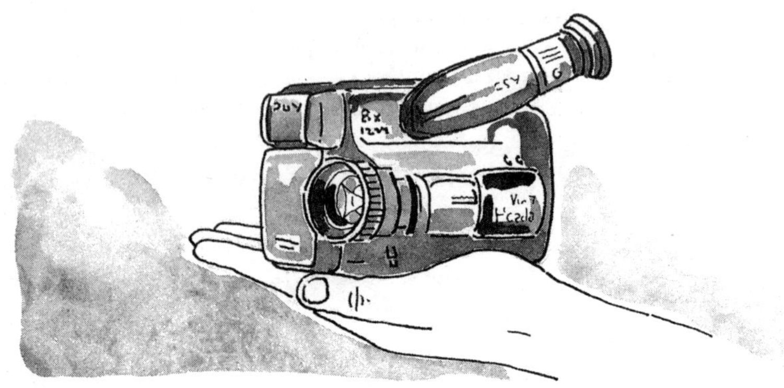

A typical palmcorder

these Palmcorders is quickly making them the most popular type of camcorder – so much so that the range of choice in the bigger models is diminishing, as manufacturers concentrate their efforts on palm-sized launches.

The convenience of a smaller camcorder is not one that you should dismiss lightly. Many people have bought larger camcorders, only to leave them at home when going out for the day or away on holiday because they cannot cope with lugging another bag around with them. A smaller camcorder can be slung in the corner of a bag you are already taking. The great advantage of the palm-sized type, therefore, is that because it is so small you might use it more often – after all, it's an expensive piece of kit to leave in a cupboard, just to be brought out for special occasions.

It is worth noting at this point that manufacturers tend to quote the weight of their camcorders exclusive of batteries and tape. As you are never going to use them without either you'll have to allow for this yourself. A rechargeable battery is especially heavy – weighing between 200g and 400g, depending on its capacity. A tape can add a further 25–250g – depending on its format and length.

It is not just the weight of the camcorder that is important. Its size is also likely to have a direct relationship on the size of the controls. The smaller the buttons, the more fiddly they are to use. How difficult is likely to depend on the size and length of your fingers. What are well-spaced, easy-to-reach controls in one person's hands can be infuriatingly small in another's.

The lesson here is that a camcorder is like a glove – it should fit the size of hand that it is intended for. When buying a camcorder it is important to try the models that interest you for size – checking that the main controls (such as zoom rocker, record button and manual focus override) fall readily to finger. Then check that the lesser used controls do not have buttons so small that your fingers cannot operate them.

Left-handers will find it particularly difficult to find a camcorder that is comfortable in their hands. Nearly all camcorders that have been made to date have been ergonomically designed with the right-handed majority in mind. Handgrips are therefore found almost exclusively on the right side of the camera (as you hold it to your eye). The viewfinder is also frequently found on the right side – and assumes that you will be using your right eye to watch what you are recording. People with a dominant left eye may find it difficult to use these camcorders – and may be forced to choosing a camcorder with a centrally placed viewfinder.

If you cannot find a camcorder immediately that feels completely comfortable don't give up hope. The three shapes of camcorder described above are the main types. Many different designs are possible, because, as the signals within the camera are passed around electronically, the shape in which the components are arranged behind the lens is unimportant. Flat binocular-shaped designs, camcorders that look like SLR stills cameras, ones that use the tall and thin cine camera design – all these and more have been sold before. Their peculiar ergonomics might be just what you're looking for . . .

While you are shopping for your camcorder, you are also likely to notice that some camcorders will look exactly the same as another, except, that is, in name. It is common practice for major manufacturers such as Sony, Panasonic and JVC to make camcorders for other companies, who do not make their own camcorders. Often referred to as 'cloning' or 'badging', this process means that a popular model can be available in half a dozen different guises. Small differences in finish, colour, warranty length and accessories mean that you might find a 'clone' a more attractive proposition than the original.

Deciding on size and shape is just the beginning to buying a camcorder. The most difficult decisions you will have to make are on the tape format that your camcorder will use and the features you need. In the next three chapters we shall look in detail at what is on offer in each of these areas – and how to decide what is right for you.

2
—— TAPE FORMATS ——

Format wars

There is nothing quite like a format war for putting you off buying the latest electronic gadget. Those who were enticed into buying Beta video recorders or 8-track cartridges will be only too keen to tell you how rushing headfirst to buy a new piece of technology can lead to long-term disappointment, if not disaster.

No one wants to buy a camcorder that is going to become obsolete in five years' time. And when you find out that there are six different tape formats commonly used in today's camcorders, you would be forgiven for quaking in your boots at having to decide which one to go for.

But before you start reaching for the Valium, I should quickly point out that choosing any one of the current video formats is very unlikely to have any long-term dire consequences. The difference between the VCR format war of the late 1970s (when VHS annihilated Beta) and today's perplexing choice is greater than it might seem.

To explain what I mean, let's look at the ways in which we have listened to music over the years and the format wars it has raised. First 78s versus 45s, musicassette against 8-track cartridge, LP against CD, MiniDisc versus DCC . . . The reason why all these choices were, and are, so important is software. Whatever your preferences or prejudices you just cannot buy today's chart busters on 8-track cartridge or 78rpm record. The record stores don't want to stock their vast catalogue of titles in a whole range of different formats – so the weakest formats disappear from the shelves, leaving the unfortunate consumer with an obsolete piece of equipment.

With camcorders the only software that is required is blank tape. You do not use them for playing pre-recorded movies or music. As long as there is blank tape available, you can carry on using your camcorder without even wasting a moment's thought on what formats other camcorder owners are using. Selling six different types of different blank tape is far easier for the stores to cope with in comparison with stocking thousands of different pre-recorded movies or albums.

Choosing a format

Despite the fact that the video formats used on today's camcorders can be used in isolation, there are important differences between the six types currently used. Each has its strengths and weaknesses and different degrees of compatibility with the other formats. To complicate matters still further, there are important differences even within models of the same format – particularly in the way that sound is recorded. The sound as recorded on one VHS-C camcorder, for instance, may not be able to be played back in full detail by another VHS-C camcorder – because one records and replays sound in stereo, while the other cannot.

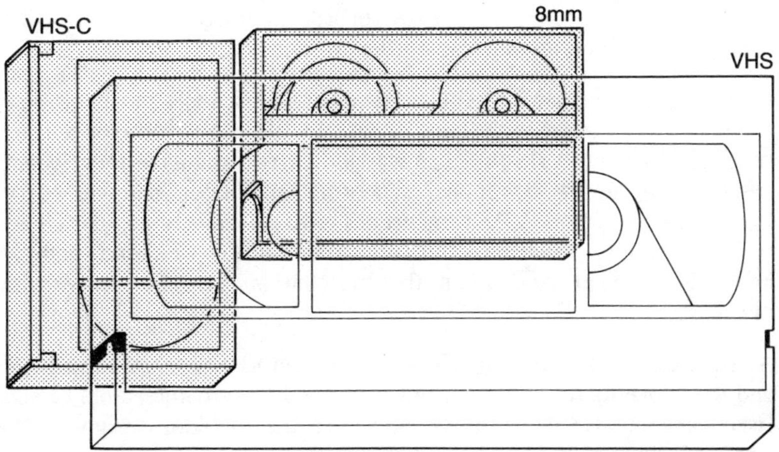

The relative sizes of VHS, 8mm and VHS-C cassettes

VHS

The VHS format (Video Home System) was invented by JVC in the mid-1970s for use on video recorders. Today, it is by far the most popular format used for VCRs in the world. It might seem logical, therefore, to presume that VHS should be the most popular format for camcorders. Although it enjoyed some early success, especially in America, it is now very much in decline as a camcorder format. As the tapes are large – almost the size of this book but twice as thick – so are the camcorders.

The advantage of being able to take your tapes straight out of your camcorder and put them directly into your home VCR is now spurned by most people – who prefer the portability offered by the smaller formats.

The important thing to bear in mind when choosing a format is that nearly every camcorder you can buy is not only a camera and recorder – it is also a playback unit too. There is no need to play back the tapes in a video recorder – all you need to do is to link the camcorder straight to the television. The only complication to this argument comes when you want to send your original recording to someone else to watch, who does not have a similar format playback machine. Even then it is a relatively easy procedure to copy recordings in any format on to VHS simply by connecting your camcorder to your VCR.

However, VHS remains a popular format among schools and institutions – where users might only have access to the camcorder for a short time, and by using VHS they can watch their recordings at their leisure without tying up the communal camcorder any longer than necessary. For the same reason, VHS camcorders have remained a popular choice at rental outlets – you hire the camcorder for the weekend, and can keep the original tape when the machine is returned on Monday morning.

As explained in the previous chapter, although VHS machines are large, and therefore heavy, they are not necessarily less comfortable to use. Practically all VHS camcorders use the on-the-shoulder design, so the strain of supporting them is not taken in the arms.

The other traditional advantage of the VHS camcorder has been its long tape lengths – giving you four, or even five, hours continuous recording on one tape (two hours on US machines, owing to the faster tape speed used for the NTSC television system). However, the 8mm format now offers up to four hours of continuous recording using a 120-minute tape on long-play setting (270 minutes in the US using a 135-minute cassette). VHS camcorders don't usually have this long-play facility.

Owing to the declining popularity of VHS camcorders, many of the models available can be outdated in terms of looks and specifications, as the manufacturers focus their efforts on the more popular miniature formats. However, the current choice does include both stereo and mono models.

The way in which sound is recorded on VHS camcorders (and those using the VHS-C, S-VHS-C and S-VHS formats) offers some advantages to the creative videographer over nearly all 8mm and Hi8 models. On mono models from the VHS camp, the sound is recorded completely separately from the video signal. While the picture is recorded in a series of diagonal stripes along the tape (using the helical scan system), the monaural soundtrack is recorded along the edge of the tape – in the same way as on an audio cassette.

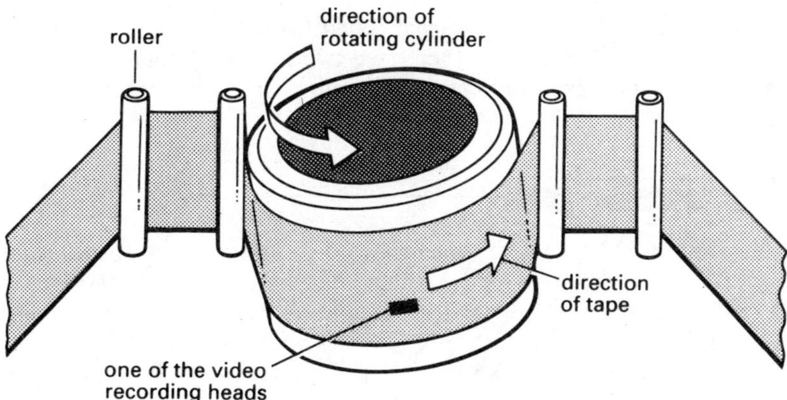

The helical scanning system

Although the quality of this sound is technically inferior to the HiFi sound of 8mm recordings, because it is completely separate from the video signal it can be replaced without affecting the picture (using a feature called audio dub). This means that you can replace the original with music or commentary in the camcorder itself. Conversely, you can re-record parts of the video without losing the originally recorded sound – this as known as insert editing.

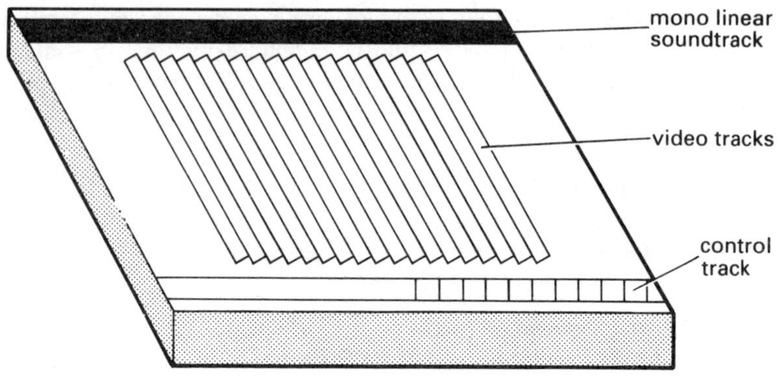

How audio and video signals are recorded on mono-aural VHS and VHS-C tapes

On camcorders with stereo sound, of whatever format, the stereo soundtrack is recorded underneath the video signals – so can not be re-recorded without re-recording the picture as well. Stereo camcorders using the four VHS format variants also record the sound on the mono linear track too – so this can be replaced using the audio dub feature, without losing the original sound (as this is still recorded on the stereo soundtrack). You then have two different soundtracks that can be replayed mixed together along with your original footage.

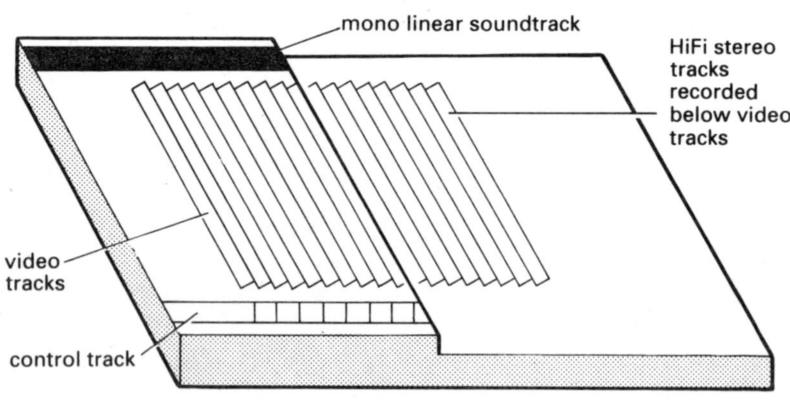

How audio and video signals are recorded on HiFi stereo VHS, VHS-C, S-VHS and S-VHS-C tapes

Although the ability to audio dub and insert edit are attractive features of the VHS formats, it should be pointed out that those who at the stage of wanting this amount of creative control are usually editing their footage anyway. Even those using an 8mm format camcorder will edit on a VHS-format video recorder – thereby gaining the ability to audio dub and insert edit.

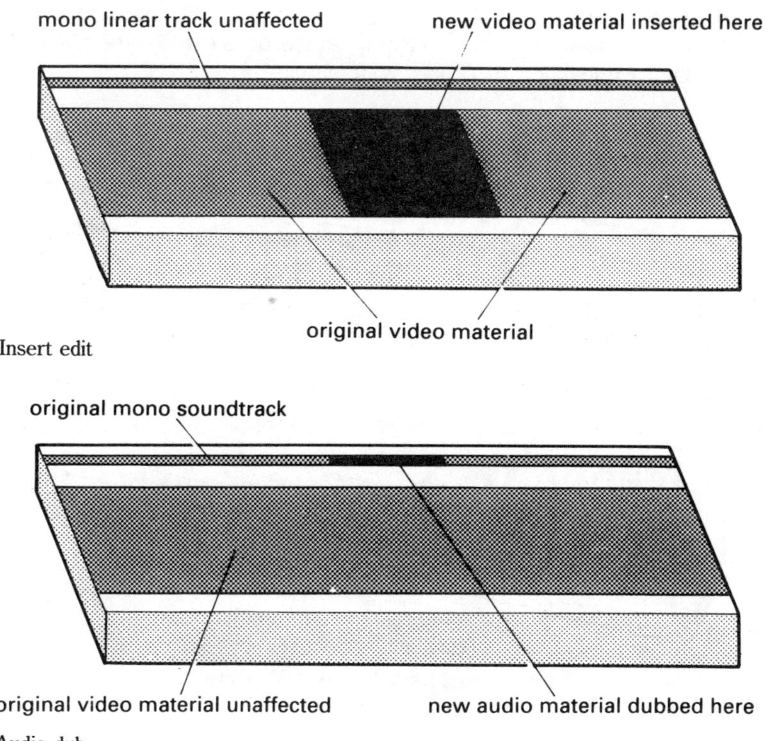

Insert edit

Audio dub

VHS-C

The VHS-C format was designed by JVC in the early 1980s to offer the compatibility advantages of VHS, along with the portability of a smaller compact cassette and camcorder (the 'C' stands for compact). In fact, VHS-C cassettes use exactly the same 1/2inch-width tape as VHS cassettes, but smaller lengths of it are housed in a smaller shell. The VHS-C cassette is about the same size as a packet of 20 king-size cigarettes.

By use of an ingenious mechanical device, called a VHS-C adaptor,

VHS-C tapes can be replayed in a full-size VHS video recorder. The adaptor is the same size and shape as a VHS tape, and has a compartment in which the VHS-C cassette is inserted. When the compartment is closed, two battery-powered arms pull out the tape to the required width and position for the VHS system. The adaptor can then be placed directly in your VCR, and your VHS-C tape watched like any other VHS tape. Of course, nearly all VHS-C camcorders can also be plugged directly into a television to replay your movies.

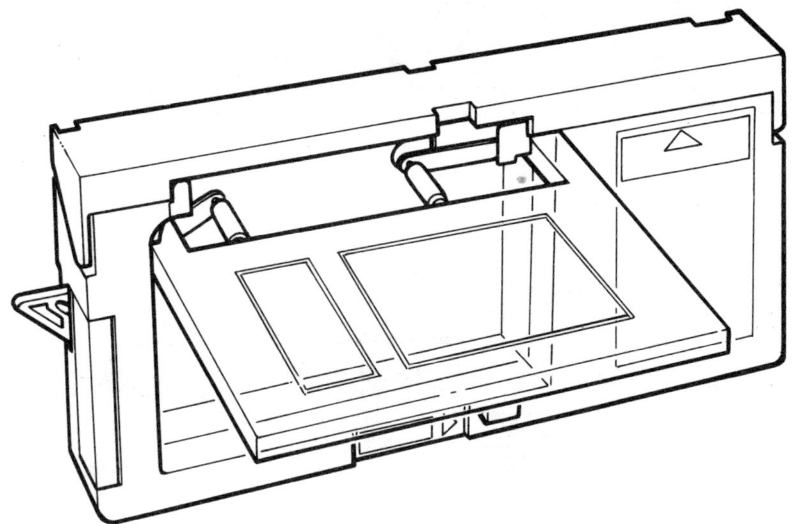

A VHS-C adaptor allows you to play VHS-C tapes in a VHS VCR

Because the format uses normal VHS tape, the maximum recording time on one cassette is not as long as some of the other formats. All the same, 45-minute cassettes are available – giving up to 90 minutes of continuous recording in long-play mode (30 minute tapes give you up to 90 minutes' recording in the US NTSC system, thanks to the Extended Play (EP) mode – which records at a third the normal speed). Please note though, that if using the long-play mode you will not be able to play back your recordings in your VCR using the VHS-C adaptor, unless your VCR also has this long-play mode.

Like its older, full-size counterpart, VHS-C camcorders are available in mono and stereo versions – with the same advantages of audio dub and insert edit (should the camcorder have these facilities).

8mm

The 8mm (or Video 8) format was designed by Sony in the early 1980s as the bespoke format for the portable camcorder. Using a special metallic 8mm-wide tape (1/3in), the cassettes are the same size as an audio cassette – making it the smallest of the formats, and therefore giving its camcorders a slight advantage in compactness. The longest tapes currently available last 120 minutes in standard play – or four hours in long play (135 minutes and 270 minutes, respectively, in the US and other countries using the NTSC television system).

Despite the fact that 8mm tapes cannot be played back in a VHS VCR, the 8mm format has gradually become the most popular camcorder format in most countries in the world. This has not strictly come about because of its advantages as a format per se, but because of more innovative and faster product development by the camcorder manufacturers in the 8mm camp.

Although a small number of 8mm video recorders are available, most 8mm camcorder owners use the camcorder itself to play their recordings back on their televisions.

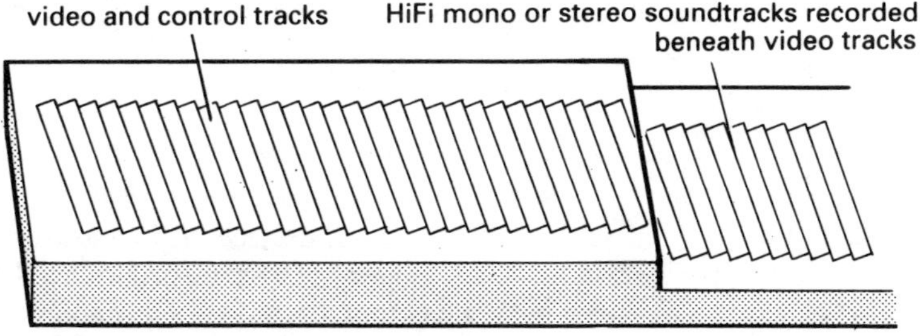

How audio and video signals are recorded on 8mm and Hi8 tapes

Whether you choose a stereo or mono 8mm camcorder, the soundtrack is essentially recorded in the same way. On all 8mm camcorders the sound is recorded with the picture using the helical scan system. This means that an 8mm mono soundtrack is written at a much faster speed than a VHS-C mono soundtrack – and because of this is of a much higher quality. In fact, the frequency response of an 8mm mono soundtrack is good enough to quality as HiFi. However, the drawback of this system is that the sound cannot be replaced independently of the video signal – so audio dubbing and insert editing are not possible in the camcorder (although they can be performed later while editing on to VHS).

High-band formats

The three so-called high-band formats were introduced in the late 1980s to offer a vastly superior picture quality over the low-band VHS, VHS-C and 8mm formats. Typically, the high-band formats – S-VHS, S-VHS-C, and Hi8 – can deliver pictures with about 400 horizontal lines of resolution, compared with an average of just 240 lines for the three low-band formats. In essence, you get a far sharper, more detailed picture.

In order to give this vastly superior picture, high-band camcorders use superior, higher-priced tapes. The other trick they use is to record the brightness and colour of the picture separately – which means far better colour clarity, and the elimination of flicker in finely striped detail. In order to take full advantage of this extra quality, you will ideally need to have a television with a special S-video socket – that allows the camera to send the brightness and colour parts of the signal separately (or luminance and chrominance to give them their technical terms).

One of the major advantages of the high-band formats is found when it comes to editing. Editing involves copying from one tape to another – and this always means a drop in picture quality. By starting off with a higher-quality picture to begin with, you are therefore less likely to be worried by the quality loss during the copying process.

S-VHS

The high-band equivalent of VHS – using the same size cassettes but using a higher quality tape. S-VHS (Super VHS) camcorders are large – but their on-the-shoulder design has found some favour among semi-professionals.

Although S-VHS tapes are the same size as VHS ones, it is not possible to play them on an ordinary VHS VCR. Special S-VHS VCRs are available, although they are primarily intended for use in editing. A few VHS VCRs are now being made with a Quasi-S-VHS function – allowing you to play back S-VHS tapes, but at normal VHS picture quality.

It is, however, possible to play normal VHS tapes on an S-VHS camcorder or VCR. In fact, you can use an S-VHS camcorder to make normal VHS recordings (on S-VHS or VHS tape) – by flicking the appropriate switch.

The maximum tape length obtainable is four hours (two hours using the US NTSC system) – and it is therefore unusual for S-VHS camcord-ers to have a long-play mode. Most modern S-VHS camcorders have stereo sound, as well as the re-recordable mono linear track – although some models exist with just mono sound alone. Audio dub and insert edit are therefore very common features on these machines.

S-VHS-C

S-VHS-C is the high-band equivalent of VHS-C – and the compact version of S-VHS. As such its camcorders usually come with a VHS-C adaptor – but this is only of any use if:

- you have an S-VHS VCR;
- you have a VHS VCR with Quasi-S-VHS playback; or
- you have made a VHS-C recording using your S-VHS-C camcorder.

The maximum recording time possible with this format of camcorder is 90 minutes – using a 45-minute tape in long-play mode (30 minutes, tripling to 90 minutes in extended play, with the US NTSC system). Most (but not all) S-VHS-C models have HiFi stereo soundtracks, as well as the re-dubbable mono linear track.

Hi8

As its name suggests, Hi8 is the high-band equivalent of the 8mm format. The tapes are the same size and they have the same maximum recording times. However, they are not interchangeable. Although you can play 8mm tapes in a Hi8 camcorder, the reverse is not usually true. The only times when you can play a tape from a Hi8 camcorder in a normal 8mm machine are if:

- the machine has a Quasi-Hi8 facility – which allows you to replay Hi8 recordings, albeit in low-band quality; or
- you have switched your camcorder to record in 8mm (such a facility is available on nearly all Hi8 camcorders).

At present only a handful of Hi8 VCRs exist – all made by Sony, and primarily intended for use in professional or semi-professional edit setups. Of course, as with all the formats, it is easy to plug your camcorder directly into your TV – and simple enough to make VHS copies of your recordings.

Nearly all Hi8 camcorders only have FM stereo HiFi soundtracks – which are recorded inseparably from the video tracks. True insert edits and audio dubs are therefore not usually possible without making a copy of your recording – and adding the new sounds or pictures at this stage.

The exceptions to this are the one or two camcorders (made, again, by Sony) that feature a PCM digital stereo track – along with the normal stereo tracks. This extra track is not recorded along the edge of the tape (as with the VHS formats) – but on the end of the normal diagonal stripes that contain the video and sound information. As the PCM sound is distinct from the other information, it can be re-dubbed at a later stage – without affecting the video picture or the standard stereo soundtrack. So audio dubs are possible (but not insert edits). What's more this PCM soundtrack is of the highest quality – being recorded digitally (like a compact disc) – and, unlike the re-dubbable track on the VHS formats, it is in stereo. Needless to say, the couple of machines with this feature are the most expensive consumer camcorders that money can buy.

Future formats

At present, all six of the available camcorder formats record information in an analogue form – they magnetise the information they receive in increasing strengths, in direct proportion to the strength of the signal. Audio cassettes and LPs record music in an analogue form. However, most of us are now familiar with the compact disc – which records its music digitally. Instead of recording information in a way that mimics the way we perceive it, a digital system converts the information into a binary code – a series of ones and noughts that are stored away in vast streams, ready to be converted back into analogue information when required.

The advantage of digital recordings over analogue recordings, as those with CD players will testify, is that they are free from background noise. What is more you will always get a perfect copy when copying a digital recording. The advantages of using a digital recording system for video are clear. Using today's video technology, it is not possible to make a fourth-generation copy (a copy of a copy of a copy of the original) without serious degradation of the picture. With digital technology it is theoretically possible to make a 94th-generation copy without any loss of quality.

Digital video for consumer use is still a number of years off. The technology is available for professional broadcasting, but is very expensive. Domestic machines are not yet possible simply because the vast amount of information that has to be stored to be able to show, what is in effect, 25 pictures a second, plus sound, for up to four hours. Currently, the manufacturers, are working on cunning ways to help simplify the information recorded, without sacrificing the result.

One thing about digital video that is almost certain is that it will still be recorded on tape, similar to that which we use today. Computer chips or small compact discs cannot currently store enough information to make their use as recording mediums for camcorders possible.

Although, digital video will mean new video formats, the existing ones are likely to last for the foreseeable future – as this new technology is going to be prohibitively expensive for most people for the first few years.

Types of tape

When it comes to choosing the tape you put in your camcorder, there is a lot of truth in the saying 'put rubbish in, get rubbish out'. Your recordings can only be as good as the tape they are recorded on.

Put simply, VHS video tape is just a strip of plastic covered in ferric oxide – that's right, rust! In the same terms, 8mm tape is just a strip of plastic covered with metal filings. However, not all tapes are made to the same standards.

Thankfully, most camcorder tapes are of a high enough standard not to worry that your recordings will be completely unacceptable because of the tape you have bought. The one exception to this is VHS tape – all manner of manufacturers with dubious credentials make VHS tape, hoping to con the innocent into buying it with its incredibly low price. It might suffice for recording a TV programme while you are out of the house – but it is hardly the stuff you want to trust your priceless video memories to. Fortunately, these cowboys have so far kept away from making VHS-C, 8mm and high-band tapes.

Nevertheless, there are a bewildering number of types of tapes on the market for the camcorder user – from the camcorder manufacturers and tape specialists alike. The difference between the cheapest and most expensive 'grades' of a particular format can be a fourfold increase in price. With names such as Master, Library, HiFi, Pro and High Grade it is not always easy to see which is meant to be best. What is worse is that their is no standardisation in terminology – so High Grade can mean the most basic grade there is. Unfortunately, the only simple to tell the difference between the grades is by price within a particular manufacturer's range.

A tape must perform a number of seemingly different functions – and recording picture and sound in the highest possible quality is only the start of it. Tape must be incredibly durable to withstand the rigours of not just being shuttled forwards and backwards through the transport mechanism – it must also stand up to the video recording heads rubbing up against it at speeds of 1500 revolutions per minute. The tape surface must be smooth enough to ensure there is not too much friction – but on the other hand, it must be abrasive enough to keep good contact with the recording heads, to protect the tape from wear, and to keep the recording heads clean. Another high priority is that the tape should not stretch – as this could seriously affect playback quality.

All four VHS format tapes are basically made in the same way. The base is coated with a binder material that contains minute particles of magnetic material. The most basic grades use ferric oxide as this magnetic material – the higher the grade, the more sophisticated the chemistry becomes. Materials such as titanium monoxide, barium ferrite and cobalt magnetite are used to increase the quality of the recordings it can make. The better the grade, the finer are the particles and the more closely are they packed together. The quality of the base film and the binder increases too. The highest grades of tape are reserved for the S-VHS and S-VHS-C formats, where more information has to be stored.

However, 8mm and Hi8 tapes are in a different league from these VHS tapes when it comes to quality. But they need to be – measuring only 8mm across, rather than 12.5mm, they have to pack the information into a much smaller space.

Basically, 8mm tapes and the cheaper Hi8 formulations are made in a similar way to the VHS formats, but instead of using metal oxide they use pure metal. For this reason, they are known as metal particle (MP) tapes. Normally an alloy of iron, cobalt and nickel is used to make the superfine particles that are held on the base film by the binder. Because the metal itself is unstable, each particle is coated with a very thin layer of oxide. Without the amount of magnetic energy that metal particle tape can store when compared with the VHS formats, the compact 8mm format would not have been possible.

In pioneering the Hi8 format, Sony took the amount of magnetic energy a tape could store a stage further. The process they came up with was metal evaporation. Instead of using a binder to hold the metal particles, a vacuum evaporation process is used to heat the metal to such a temperature that it turns to gas which condenses directly on the base. Without the binder, the metal particles are therefore packed as closely as they can be on the base film – allowing them to hold more picture detail and information about colour. Metal evaporated (ME) tape is recommended by the camcorder manufacturers as the ideal choice for Hi8 recordings – unfortunately it is relatively expensive, and for many, metal particle tape has to do.

—————— Looking after tape ——————

Because video is such a new medium, no one can honestly say how long a video tape will last without its quality deteriorating. VHS tape was first produced in the 1970s, and 8mm tape in the 1980s – hardly enough time to tell whether video recordings really stay the test of time. However, we do know that video tape is a fragile substance – and needs proper care if you are going to enjoy the memories it holds in years to come.

There are two different ways the longevity of tape can be measured. The first is how many times a particular tape can be re-recorded on before there is any deterioration in the quality of the recordings. The more important is how long a particular recording on a tape can last before there is any quality loss. Unfortunately, it is only on the first count that video tape excels – you can record over your tapes hundreds of times without problems. This is ideal for recording television pro- grammes when you are out, but not so useful for home movies that you will want to keep for the rest of your life.

The trouble with the recordings themselves is that each time you play them back there is a slight signal loss. The first point, therefore, with any important recording, is to try and minimise the number of times you play it back. If you know that you'll want to play a tape again and again, make a copy of it and watch this, keeping the master recording safe for when you really need it.

It is also known that high humidity and high temperature accelerate the ageing process. It is suggested by the manufacturers that it is best to store tape between 15–20°C and between 40–60% relative humidity. Most importantly you should avoid keeping tapes where there are great fluctuations in temperature. Avoid keeping them in a place where they will receive direct sunlight – or in an attic where they will get very hot in summer and very cold in winter. Keep them away from kitchens and bathrooms where there is likely to be excessive moisture. Putting tapes near to magnetic sources – such as on top of a television, other electrical devices or magnetic children's toys – is also taboo.

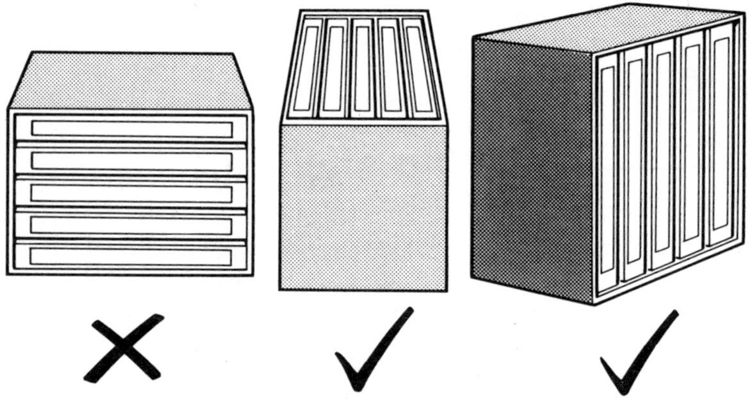

Never store your tapes flat. Always store tapes on one of their edges or on their ends

Another essential point is that tapes should never be laid flat – they should always be stored on either of their ends, like books in a library, or on their sides. The protective cases or sleeves that come with the cassettes should always be used – plastic cases that click shut are best as they minimise dust getting to the tape. Also, before putting your tapes away you should rewind them to the beginning.

For tapes that are going to be stored for long periods without being played, it is recommended that they are fast-forwarded and rewound periodically, to ensure the tape is ventilated. Sony recommends that this is done at least every three years.

The most serious of tape hazards is human error – and that's not just the risk of a tape falling in a cup of coffee or being accidentally thrown out with the rubbish. The most likely accident is that you'll record over your priceless footage. You should therefore quickly familiarise yourself with the record protect systems that are built into every tape. On the four VHS format tapes, there is a black plastic tab, which when broken off will ensure that any VCR or camcorder will not record over it (if you decide later to re-record over it, this fail safe device can be defeated by sticking tape over the hole that is left when the tab is removed). However, 8mm and Hi8 tapes use a slightly different system. Instead of a breakable tab, there is a slidable one which is normally coloured red. Confusingly perhaps, when the tab is pulled across the hole you cannot record, and when the hole is left open you are free to record.

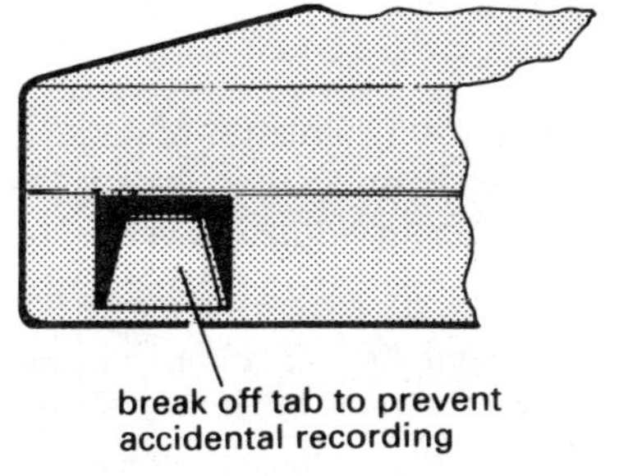

VHS formats

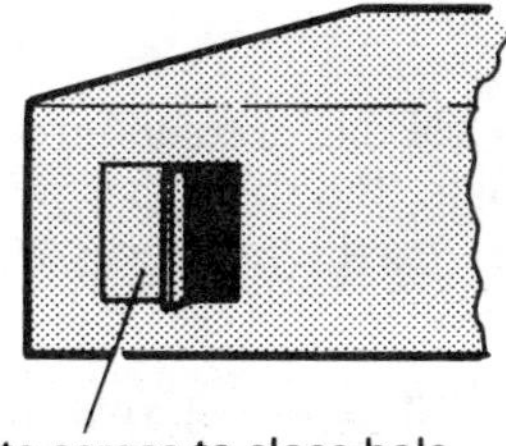

8mm formats

Finally, for those very special recordings, it is a good insurance policy to copy them every five to ten years. If your tapes do suffer deterioration, you will then have a good copy to take over as the new master recording!

If all this advice sounds like scare-mongering, it is meant to. In 20 years' time it might be too late to complain that you cannot show your footage of your children to your grandchildren – you must start looking after your recordings right from the very start.

3
— BASIC CAMCORDER — FEATURES

While the size and format of a camcorder is largely a matter of personal taste, the thing that really separates the vast array of camcorders that are on the market is the range of different features that each of them has.

When choosing your first camcorder it might be difficult to imagine which of those little buttons is really going to come in handy. No one wants a whole bank of controls that you are never going to touch. But while some features are essential, others are little more than cheap marketing gimmicks. It is important that you can tell the difference between the two if you are going to make an intelligent choice – especially as the stores and the manufacturers' brochures will sing the praises

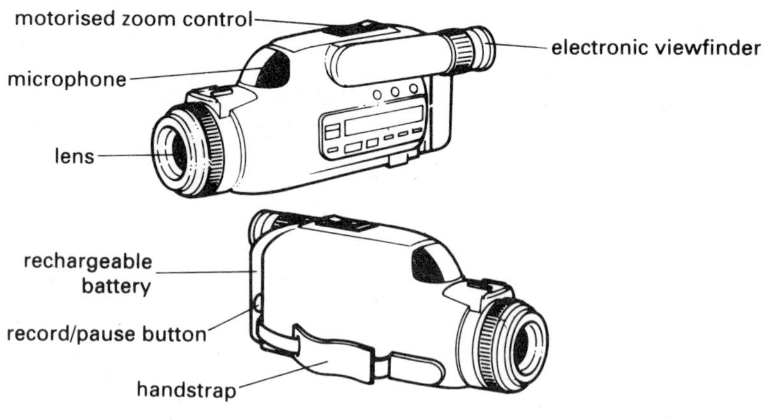

The main features on a camcorder

of the latest fad, rather than pointing out the features that have been stripped off to save costs

For those who are new to the camcorder game, it is important to realise that there are some general features that are shared by every single camcorder on the market. These, mostly, are the essentials, but still include buttons that you might be able to do without if you were given the choice. The others, the ones that differ between model to model, are either those that are useful, or even essential, for true creative movie-making, or they are the gizmos, some of which are useful inventions, others which are a waste of the circuit boards that they are printed on.

This chapter, and the chapter that follows, go over every feature you are likely to come across in detail. Not only will the function of each of these features be fully explained, its real use for the videomaker will be assessed. In the rest of this chapter, the general features you'll find on every camcorder are listed and explained. The optional features that are only found on some camcorders are dealt with in Chapter 4.

Zoom lenses

There was a time when only some camcorders had zoom lenses, but today they come as standard on practically all models. On 99% of them the zoom is built-in – but there are a few where the lens or part of the lens is interchangeable to give you a greater choice of optics.

The primary purpose of a zoom lens is that it allows you to get 'closer' to your subject without having to move yourself. It does this by having a variable focal length – with a wide view at one extreme, to a powerful telephoto one at the other. But the beauty of a zoom lens is that it gives you all the possible focal lengths in between – giving you complete control over what, and what not, to include in the picture. Typically, the widest setting on a camcorder's zoom might give you a 45° angle of vision, and the most telephoto lens would restrict your view to just a 6° arc.

All of today's zooms have a motorised control to shift between the wide and tele settings. This motorised control allows you to zoom in on a subject smoothly during recording. However, as we will see in later chapters, it is not good practice to zoom during recording. Rather, the zoom control is primarily used for framing up a shot before recording – thus a manual zoom control is useful, as it helps cut down on battery power consumption.

The power of a particular zoom lens is commonly measured by its zoom ratio – a figure that compares the degree of magnification between the widest and most telephoto settings on the zoom. Thus on a lens with a zoom ratio of 6x, the focal length of the telephoto end is six times larger than at the widest end. To put it another way, the wideangle end is six times wider than the telephoto end.

Common zoom ratios available are 6x, 8x, 10x and 12x. These figures should not be confused with those given for electronic zoom ratios. A

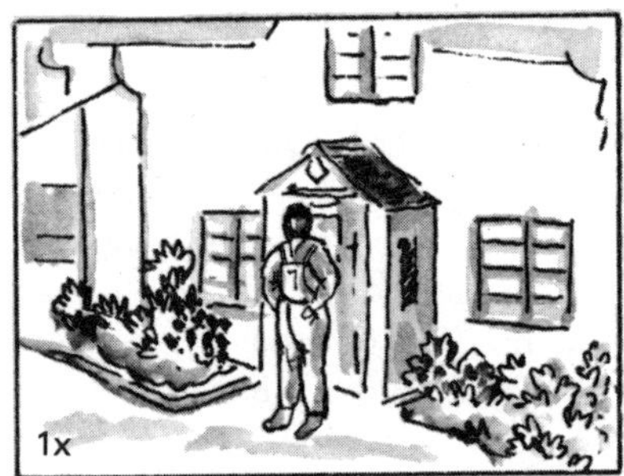

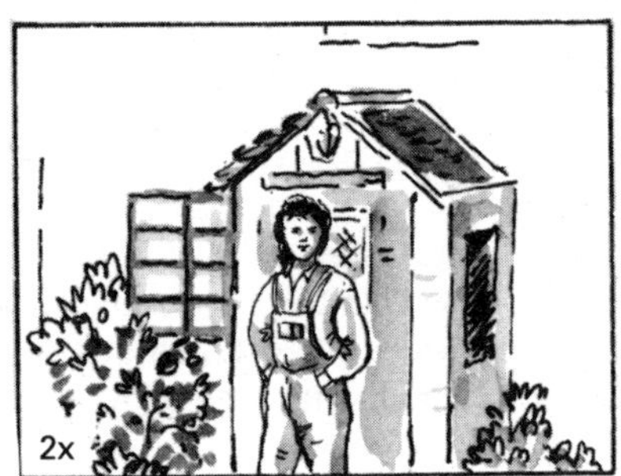

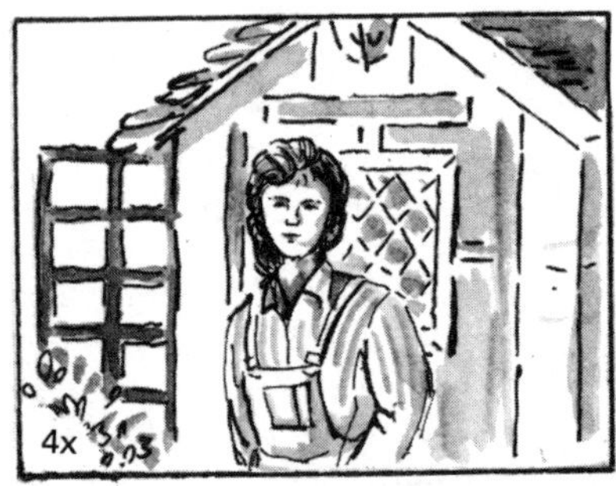

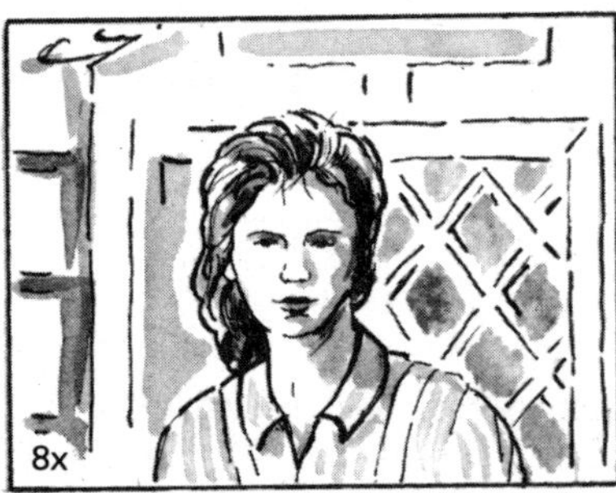

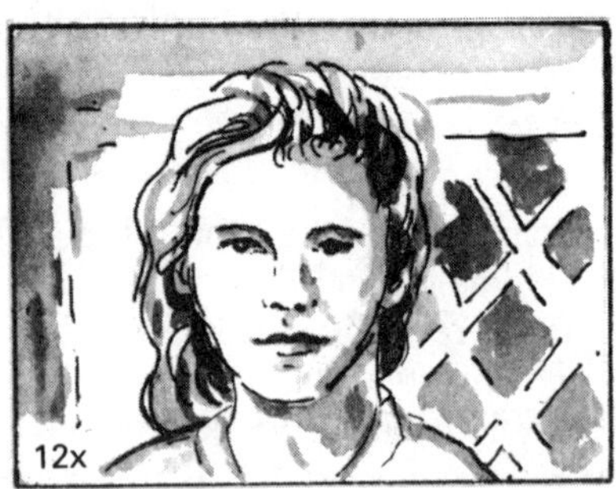

Zoom range

camcorder which boasts a 100x zoom ratio has not really got a 100x zoom lens, rather it has a 10x zoom lens, but also has an electronic digital device that can enlarge the central part of the picture to give the appearance of 100x magnification – however the picture quality is vastly inferior (see next chapter).

Having a larger zoom ratio is not necessarily better. In comparison with stills cameras, a 6x zoom still affords an extremely powerful telephoto setting – suitable for most applications, including sport. The more powerful zoom ratios do have some appeal for wildlife – especially if your prey is small and hard to approach (as birds are, for instance). The drawback, is the longer the telephoto setting, the harder it is to hold the camera steady enough to get a stable picture.

It is, perhaps, more useful to look closely at the wideangle end of a zoom when choosing a camcorder – rather than automatically picking the one with the biggest zoom ratio. The wideangle settings on camcorders are not all the same – some give a much wider view than others, and this can be more useful in day-to-day video making than a powerful telephoto. When indoors it is useful to have the widest angle of view available so that you can fit as much of a small room into the shot as possible. Outdoors a wider lens is essential for landscape shots or for squeezing large buildings into a single frame.

Working out which camcorder has the widest angle of view is not simple. The thing to look out for is the figures quoted for the zoom range – a typical example would be a 10x zoom with a focal range of 6.1–61mm, with the widest setting therefore having a focal length of 6.1mm. However, it would not necessarily be the case that another camcorder, with a focal length of 5mm, would have the wider angle of view. This is because the magnification doesn't just depend on focal length – it depends on the size of the CCD imaging chip too. The CCD chip is the device at the back of the lens that converts the light patterns that make up a picture into electrical signals. If the chip is smaller, a shorter focal length of lens is needed to give a wider angle of view. Most camcorders now have a 1/3in CCD, although 1/2in chips are still relatively common. A handful of semi-professional cameras use a 2/3in CCD – and 1/4in chips have just been developed for use in palm-sized models. The most common way to compare different camcorder zooms is to convert the focal length figures to what they would be equivalent to on a 35mm stills camera. For a camcorder with a 1/3in CCD, you multiply its focal lengths by 7.25, for one with a 1/2in chip you would multiply by 4.8, and for a 2/3in chip you would use 3.6 as your figure.

Using these equations if you end up with an equivalent focal length of 35mm, then the camcorder has a very respectable wideangle lens. The following table will help you make at-a-glance comparisons:

Comparison of focal lengths

35mm stills camera	Camcorder with 1/4in CCD	Camcorder with 1/3in CCD	Camcorder with 1/2in CCD	Camcorder with 2/3in CCD
Wideangle				
28mm	2.9mm	3.9mm	5.8mm	7.8mm
35mm	3.6mm	4.8mm	7.2mm	9.6mm
40mm	4.1mm	5.5mm	8.2mm	11mm
50mm	5.2mm	6.9mm	10.4mm	13.8mm
60mm	6.2mm	8.3mm	12.4mm	16.6mm
Telephoto				
200mm	21mm	28mm	42mm	56mm
300mm	31mm	41mm	62mm	82mm
400mm	42mm	55mm	82mm	110mm
500mm	52mm	69mm	104mm	138mm
600mm	62mm	83mm	124mm	166mm

CCD chip

The CCD chip is the heart of the modern-day camcorder. It is this ingenious little device that translates light into electrical signals. It is alternatively known as the pick-up or imaging chip. The CCD (charge coupled device) is made up of thousands of little cells called pixels. These are arranged in a grid formation, and each generates its own electrical signal when light reaches its surface. Typically, a CCD will have 300,000 or more pixels – and up to 470,000 for a high-band camcorder.

As mentioned before, there are four common sizes of CCD chip. It used to be a general rule that the larger the chip the better the picture, but on-going improvements to the manufacture of these devices mean that more and more camcorders are using the smallest 1/3in (or even 1/4in) CCD chip. The reason for this is that the detail in a video picture can only reach a certain point – dictated by the tape format used. High-band formats cannot record more than 410 lines of resolution – while the low-band formats (VHS, VHS-C and 8mm) can only resolve 250 lines. Improvements to the CCD chip cannot improve this theoretical maximum.

Confusingly, some manufacturers may claim that their Hi8 camcorder can resolve, say, 600 lines – but this only means the camera section of the machine is capable of distinguishing this much detail, it cannot possibly be made real use of.

A new trend among top-end camcorders is to use two or three chips to improve the quality. Although a 3CCD camcorder still cannot beat the theoretical maximum possible for its format, it does give other advantages that help picture performance.

Autofocus

However good the rest of your camcorder, out-of-focus footage is going to ruin your movie. To avoid this embarrassing state of affairs, every camcorder has an autofocus system – the sole function of which is to ensure that the lens is focused on the subject, so that the recorded image is sharp. The difficulty with any autofocus system is that it can sometimes focus on the wrong part of the image – or refuse to focus quickly enough for you to get the shot you want. Autofocus, in short, in not 100% foolproof.

There are two different types of autofocus (AF) system commonly used in camcorders – active and passive. The first actively tries to calculate the distance between the camera and the subject. The second, more passively, studies the image, sees how sharp it is, and then moves the lens as necessary. Both systems have their advantages - and their pitfalls.

The active system is more commonly known as infrared autofocus. The camera fires a beam of infrared light at the subject and waits for it to bounce back. Measuring the angle of the reflected beam it can calculate the distance – as the closer the subject the wider the angle. Its great advantage as a system is that it can work in absolute darkness, but its limitation is that it is less accurate for distances beyond 10 metres from the camera, making it less suitable for telephoto work.

The passive systems are also known as through-the-lens or phase detection autofocus. Although there are different methods used, basically the AF system analyses the image formed on the CCD chip and uses areas of high contrast to calculate how much lens adjustment is needed to bring the image into sharp focus. The main advantage is that the system can be extremely accurate even with telephoto settings (which for reasons we'll see in later chapters need more accurate focusing). The downside is that the system needs reasonable light, and, more importantly, a subject with a reasonable amount of contrast variation.

As autofocus systems become more sophisticated, they offer facilities such as being able to make an 'intelligent' decision as to which object in the picture is the principal subject, or being able constantly to adjust focus for a moving subject. But however clever the AF system, there will always be times when it will fall down, and for this reason any good camcorder will give you the option of manual focus (see next chapter).

Auto exposure

Every camcorder on the market today will automatically adjust itself so that it will give you the correct exposure in a wide variety of lighting situations – from a dimly lit room to a sun-scorched tropical beach. The main way that the camera adjusts the amount of light reaching the CCD chip is the iris. Performing the same function as an aperture on a stills camera, it is made of a series of blades that form a hole that can be opened and closed to adjust the amount of light getting in. This one control caters for nearly all lighting situations, but it can't cope with extremes, so two further automatic exposure controls are also commonly used.

The first of these is called automatic gain. This system comes into its own in low light as it electronically boosts the signal being received by the CCD chip. It works well – but in boosting the picture it also increases the amount of 'noise' in the signal giving a characteristically grainy picture.

The other exposure control that can be used is to engage a higher shutter speed than normal. A camcorder always records 50 fields per second (60 fields per second in the US) and usually this is a continuous process, with each field taking 1/50th of a second to record (1/60sec in the US). Each field, in fact, is half a picture recording the alternate lines of the picture – the next field records the missing lines in between. Two fields make up a whole frame. If the shutter speed is increased, however, the amount of exposure is decreased and this can be very useful in very bright conditions – or in situations when (for reasons to be discussed in later chapters) you don't want to close the iris. However, if the shutter speed is increased too much, there are effectively gaps on the tape where nothing is recorded – as the tape is still only recording 50 frames per second. With fast moving subjects, therefore, the image can seem to jump from frame to frame as it is played back. However, the effect can be useful for slower subjects.

As with autofocus, automatic exposure cannot be faultless – and therefore it is useful to have some manual exposure controls, and preferably manual iris adjustment (see next chapter).

Backlight

The most common instance in which the autoexposure system fails to give the exposure that you want, is when your subject is standing in front of a window. Here, the subject is mainly lit from behind (or backlit) – or to put it simply, it is much brighter outside than it is indoors. The result is that the autoexposure system will tend to try to expose for the bright light. This means your recording will show clearly what you could see through the window, but your subject will be turned into a dark silhouette. To compensate for this, nearly all camcorders have a special backlight button. By pressing this, the amount of exposure is increased by a set amount – allowing you to see your subject more clearly. The down side, is that the window will be overexposed, and will appear as a white, burnt-out mass. Normally, the camera will make this correction by opening up the iris. If the iris is fully open already, it is achieved by boosting the gain. A more flexible way of dealing with these exposure problems is to have full manual control over the iris settings, but this is generally only available on the most expensive camcorders.

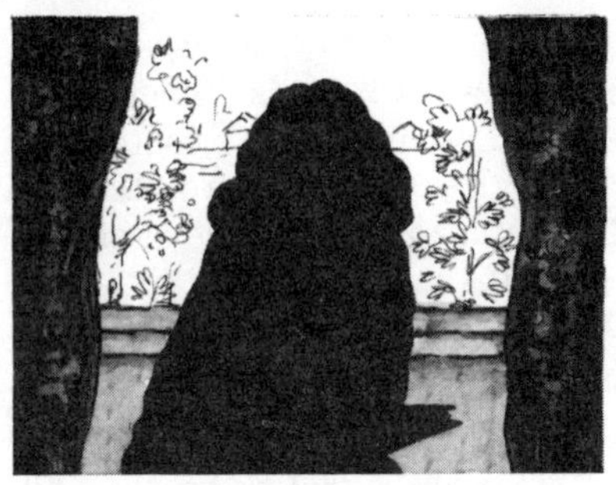

A backlight situation: the figure standing by the window appears as a silhouette

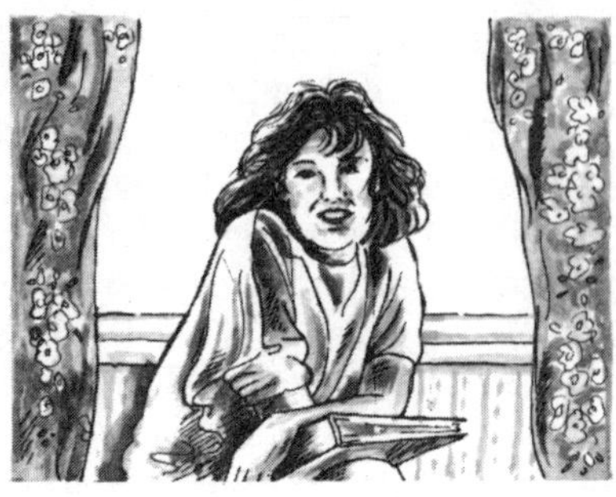

Using backlight button: the figure is now better exposed, but the view from the window is burnt out

Auto white balance

To those entering the world of camcorders for the first time, the auto white balance system is perhaps the most curious. The human eye is a remarkable piece of biological engineering, so a piece of white paper looks white whether we look at it in candlelight, in floodlights or under the midday sun. In reality, however, the colour of all these lights is slightly different – they all have different colour temperatures, to use the correct scientific terminology. Normal household bulbs give an orangy colour – called tungsten light – while fluorescent strip lighting gives a green colour cast. Even sunlight has its different colour temperatures – sunrise at one extreme is orange in colour, while a clear sky gives blue sunlight. In order for the camcorder to record colours in the way our eyes expect them to look, it has to make allowances for all these colour variations – and this is what the white balance system is for. White balance is explained in more detail in Chapter Eight of this book.

High shutter speeds

It is not uncommon for a manufacturer or a store to make a large fuss about the top shutter speed of a particular camcorder. The phenomenon has reached such ridiculous proportions that shutter speeds as fast as 1/10,000th of a second are not uncommon. Unfortunately, as good as it sounds, this is undoubtedly the least useful function commonly available on camcorders. As explained earlier, whatever shutter speed you use, the camcorder still only records at 50 fields per second (60fps on US models). So if the subject is moving at a reasonable speed and recorded using a high speed shutter setting it will appear to jump or flicker as it is played back at normal speed. If a racing car travelling at 200mph is shot using the 1/10,000sec shutter speed it will move over five feet between fields!

What the high shutter speed was designed for is to analyse movement in slow motion. So if you want to analyse your tennis serve you set up your camcorder using the high speed shutter – then watch footage in slow motion to see where you are going wrong. Unfortunately, camcorders do not normally have good slow motion facilities – so if you have an 8mm format camcorder you would have to copy the tape on VHS before analysing your action on a VCR. It is said that the Japanese obsession for golf is to blame for the proliferation of high-speed shutters on camcorders.

Low light performance

Another feature that is overplayed on camcorders is their low lux performance. One model may boast the ability to be able to shoot down to four lux, while another may seem to be able to cope, at best, with eight lux. The figures are bandied around as if they had huge significance for your movie-making. However, before you decide on the merits of a two lux camcorder against a ten lux model, it is worth pointing out that the intensity of light in a normally lit room at night would measure about 100 lux. Even candlelight would give you ten lux. In short, all camcorders are remarkably good at recording even in the lowest of light. Also it is not so much whether a camcorder can record at a particular intensity of light – but how well it can record. It is not so good if the picture is colour and the colours are murky. In truth, it is generally the most expensive camcorders that give the best results in low light – even if the quoted low lux figures do not always suggest this. In practice, if light levels are really this low you will have difficulty seeing what you are doing anyway!

Rechargeable batteries

One of the great beauties of the camcorder is that although it is expensive to buy in the first place, it is relatively cheap to run. This is due in no small part to the fact that all camcorders come with rechargeable batteries. These can be recharged from the mains at least 500 times as long as you take reasonable care of them.

The majority of camcorders use nickel cadmium batteries (*nicads*, for short) – although you may come across other types. Some of the larger models, for instance still use the older lead acid type rechargeable batteries. Most recently some camcorders have started using nickel hydride and lithium ion batteries as they offer advantages over the more usual nicads.

The reason why it is important to take special care with any rechargeable batteries is that they only provide power to your camcorder for a short period of time before running flat. Although the battery that comes with your camcorder will boast that it will last for 40 minutes or more before it needs charging – this can mean, in practice, that it might only give you ten minutes of recording. There are two reasons for this. The first is that the length of time the battery will last will depend to a large extent on how many of the motorised controls you use on the camera. The autofocus system, the power zoom, the fast forward/rewind controls all use up power, but it is possible to use the camcorder without using any of these functions. When the camera is new, it is especially tempting to play around with all these controls. The second and not unrelated reason for the shortening of battery life is that you will often have the camera switched on without making any recording. You need the power on to be able to use the electronic viewfinder on the camera – so you can see what you are planning to shoot. You then use this viewfinder while you play around with the focus and zoom controls as you frame up your shot before you press the record button. If you are not careful your battery could run down before you have committed a single shot to tape! Not surprisingly, a spare battery is one of the most useful extras that you can buy for a camcorder.

To recharge your battery, you generally unclip the battery from the camcorder, and attach it to a charger unit that plugs directly into the mains electricity supply. To make things easier for international travel, these units will usually work with any voltage from 100–250V and 50–60Hz, without any need to move any switches. The charger should be supplied as a standard accessory with your camcorder and will recharge the supplied battery in about 40 minutes (your instruction book will give precise details).

Unfortunately, rechargeable batteries, especially nicads, do have their

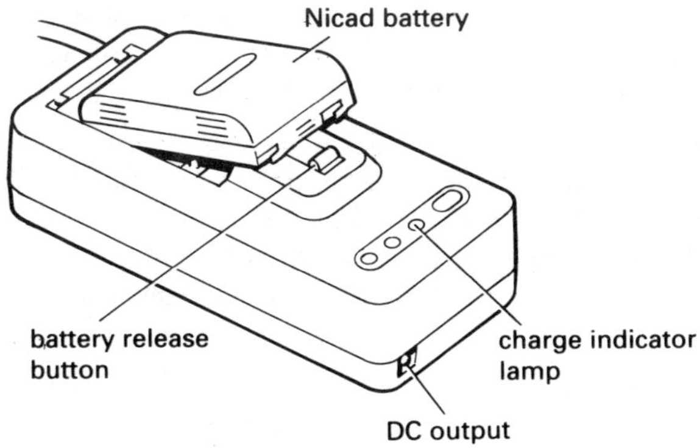

A NiCad battery charger

drawbacks. Most commonly, for no apparent reason, they can soon start to give shorter and shorter periods of power before they need recharging again. The most frequent diagnosis for this problem is called the 'memory effect' and is caused by the user continually topping up the battery before its charge has been used up. What happens is that after a while, the battery gets lazy – only some of the cells are ever properly used, the others go to sleep. To be fair, not all battery faults are caused by the memory effect, but nearly all ills are cured, or avoided, in the same way.

Nicad batteries lose their charge continuously – even if they are not being used – with one per cent of their charge being discharged every 24 hours. So there is no point recharging batteries after you have used your camcorder if it is going to sit in a cupboard for days. You should always charge your batteries as soon before use as possible.

You should also only recharge your nicads after they have run down – they should not be topped up with power. However, you don't want to go out with your camcorder knowing your batteries only have a few minutes of life left in them. So you must discharge your battery in some way before recharging it. Some chargers now have a built-in discharge or refresh mode that can be operated when necessary. Alternatively, discharger units are available as an accessory. If you have neither of these you can simply leave the camcorder on, or use the powered zoom or built-in light, until the camera automatically switches itself off. A camcorder never allows the battery to completely discharge itself of power – there is always just enough power, after automatic switch-off, to be able to eject your tape.

Whether charged or not, batteries should always be stored in some kind of case or plastic bag when not in use. This will prevent any metal object accidentally short circuiting the battery. Also it will help avoid dirt, grease and oxide building up on the battery contacts. The contacts, in any case, should be periodically cleaned with a cloth to ensure good conductivity.

It is also worth remembering that every camcorder can be run directly from the mains. This is not only useful for making long recordings indoors, but also for powering your camcorder to act as the playback unit after a day's shoot. A camcorder can be connected to the mains electricity supply via the battery charger. Either a DC-in socket is provided on the camcorder, or a special adaptor plate is used which takes the place normally occupied by the battery. In either case, the required lead should be supplied with the camcorder.

Long play

Whether your camcorder will have a long play button will largely depend on what tape format the recorder uses. On the miniature formats (8mm, VHS-C, Hi8 and S-VHS-C) the provision of a long play mode is nearly universal. On full-size VHS and S-VHS machines you will have to be content with just the standard recording speed, but as the tapes are longer to begin with this is not a problem.

The long play mode allows you to double the maximum recording time of your tape. It does this by halving the speed that the tape travels through the camcorder. It necessarily means a drop in quality in the recording – although in relative terms this is only slight. In America, VHS-C and S-VHS-C machines have an extended play (EP) mode that triples the maximum recording time possible.

VHS and VHS-C camcorder users who use their VCR to replay their videos, must ensure that their VCR also has a long play (or EP) mode if they are to use this facility on their camcorder.

As you should always aim for the highest recording quality possible, the long play mode should only be used when really necessary. It can prove so when you have to record an event in its entirety, and cannot afford to miss a few seconds as you change tapes – a stage play or a football match are good examples.

In normal situations, it is important to remember that a half-hour movie is a long movie. The secret of good videomaking is to keep things as short as possible.

Playback controls

As mentioned in the introduction to this book, a camcorder is basically a VCR with a camera attached. As such, one of the panels of controls on your camcorder will resemble those found on the front of your VCR. These are the tape transport controls, and are mainly used when it comes to playing back your tape through the viewfinder or a television. Stop, Play, Fast Forward, Rewind and Pause are all familiar enough to most of us to need no further explanation. However, to be able to use these controls it is normally necessary to switch the camcorder from camera mode to playback or VCR mode. Sometimes this switch takes the form of a panel which is pushed back to reveal the transport controls. Like a VCR, the fast forward and rewind controls often have at least two functions. If they are used when the tape is stopped they will act as normal high-speed forward and rewind. If they are used when the tape is being played, they act as picture search controls – where the tape is actually played at a faster than normal speed, so you can see what is on the tape. Confusingly, the rewind button often has a third function when you are in record pause – this gives you record review, which lets you look at the last few seconds of what you have just recorded.

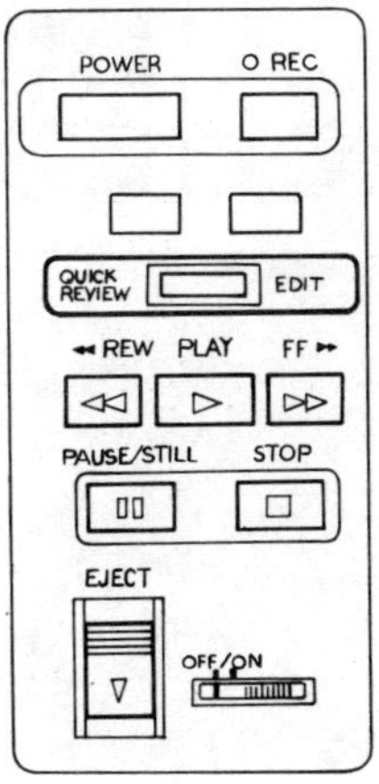

Tape transport controls

The transport controls on most camcorders are rough and ready compared with today's VCRs. They can rarely be used with complete accuracy to find the exact frame that you want. This can have its drawbacks when it comes to editing – as you will find it hard to pinpoint the exact frame you want to start on. Very occasionally, camcorders have a still advance mode, which will forward the tape a frame at a time while in pause mode – this is very useful for editing. Another drawback of camcorders is that they rarely give a good quality still frame when in pause – it is usually marred by noise bars across the picture.

Fader

The fader is one of the few creative effects that is found on nearly every camcorder that you can buy. By pressing the button, and keeping it pressed, the picture will gradually fade away to leave a plain white or black screen. This can be recorded at an end of a scene to mark, say, the end of a day. A new scene can similarly be faded in by keeping the fader button pushed in before you press the record start button. As

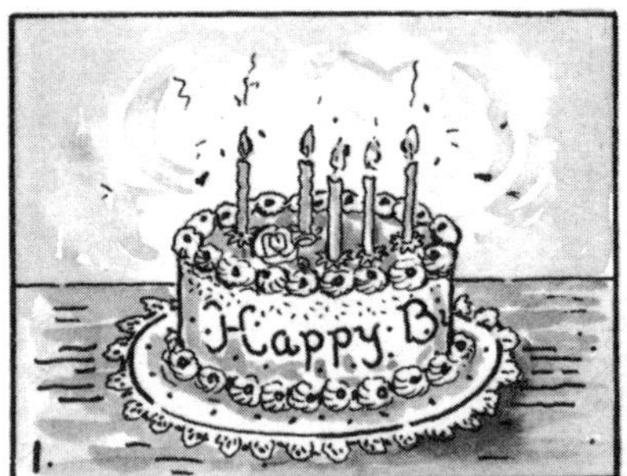

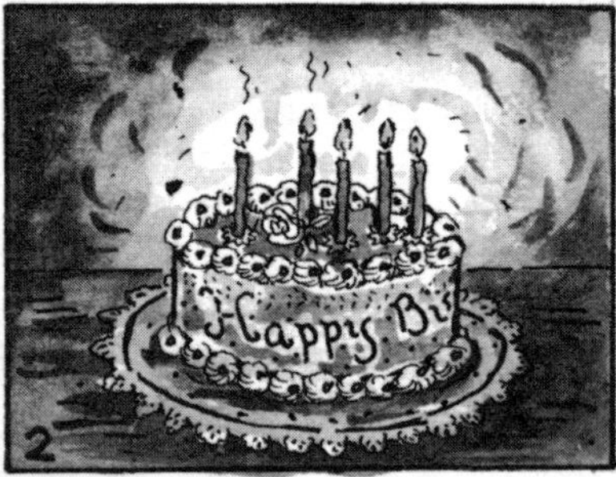

Fade to black

recording starts, you release the fader button and the picture gradually appears on the screen. It can be a useful effect – but as with any effect should be used sparingly. If you have a choice, a fade to black looks the more professional. Unfortunately, the Japanese, for some reason, prefer a fade to white - so this tends to be the more common. Some faders will fade sound as well as picture, while others will just fade the picture. Be sure you know which your camcorder has to avoid ending up with unwanted comments on your soundtrack.

Viewfinder

It comes as a surprise to most people that the viewfinders on most camcorders give a black and white picture. After all, the camera is recording in colour, so why can't you see what you are shooting in colour? Until recently, it was impossible to make a colour viewfinder that was small enough, or cheap enough to be viable. Even now, the colour viewfinders that are available usually make the camcorder £100 more expensive – and also have disadvantages over the black and white sort.

Unlike stills cameras, a camcorder's viewfinder does not give you a direct view of what the lens is seeing. Instead, the signal from the CCD chip is transferred to the eyepiece electronically – hence this device is known as an electronic viewfinder, or EVF.

Typically a camcorder viewfinder has a 0.6in black and white television tube. This is viewed through some kind of magnifying glass. On smaller camcorders, the viewfinder must be pulled out so that the magnifying glass is the correct distance from the screen. You must then, on all camcorders, adjust the distance of the magnifying glass to suit your eyesight using the knurled ring behind the rubber eyepiece. Rather than looking at the picture, it is best to make this adjustment using the on-screen display – the words and numbers which are shown in the viewfinder.

These words and numbers are a key part of the viewfinder's function. They are used to give you important information, such as whether the camera is actually recording – or is in record pause mode. Warnings are also provided, such as that battery power is low, or that moisture has got into the camera.

Instead of using a television tube, colour viewfinders use an LCD (liquid crystal display) system. This is made up of thousands of green, red

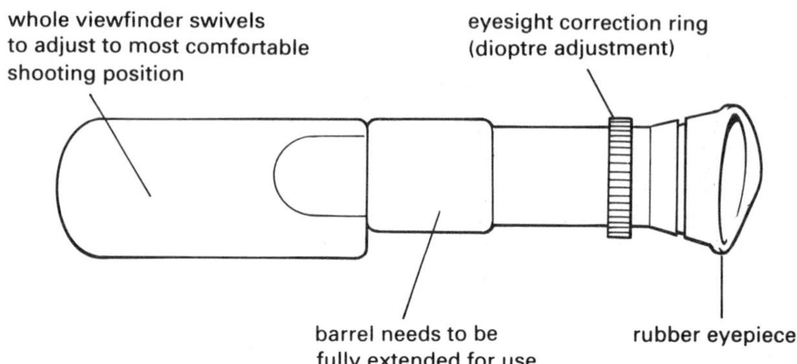

An electronic viewfinder

and blue elements (or pixels) that turn on and off to make up the picture. The disadvantage is that the picture is coarse – as you can still just make out the individual pixels – so the resolution is lower than using a black and white viewfinder. This makes it less useful for manual focusing – as it is harder to see which parts of the picture are sharp. Although, in principle, colour viewfinders should be more useful for checking white balance, in practice their colour rendition is not accurate enough to rely on.

Where colour viewfinders come into their own is for subjects where a black and white image is a hindrance to picture composition. Without colour, it is difficult to tell a football team playing in blue from a team playing in red. Similarly, insects and animals become better camouflaged when their colour is stripped away – making them difficult to find in the viewfinder. Colour LCD displays are still in their early stages of evolution, and it is very probable that before long their problems will be ironed out, and their use on camcorders will become the norm.

Microphone

The audio quality of your recordings is largely determined by the tape format you are using, and the way that the soundtracks are recorded. As described in Chapter Two, sound cannot only be recorded in stereo or mono, it can also be recorded in HiFi quality or on low-quality linear tracks. However, audio quality is also determined by the abilities of the microphone you are using.

All camcorders have some form of built-in microphone, but in absolute terms, this is one of the cheapest parts of the camcorder. As such, a built-in microphone has its limitations. For a start it is mounted on the camera, so it can be a reasonable distance away from the sound you want to record – although light can be recorded easily over great distances, sound doesn't travel very far without deteriorating. Also, it is designed to pick up sound mainly from the direction in which the camera is pointing. However, noise to the side and behind the microphone will be picked up too if it is loud enough or close enough. It is therefore possible for an on-board mic to pick-up the noise of the camcorder's motors, or the breathing of the camera operator.

More sophisticated models have zoom microphones that modify their angle of sensitivity depending on the how much the lens is zoomed – this can help minimise some of the effects mentioned above. But however sophisticated it is, a built-in microphone is never going to be ideal in all situations. For this reason, the best camcorders have a mic socket allowing you to plug in a separate microphone, which can not only be chosen for the task in hand, but can also be positioned best to pick up the sound required.

Date function

Nearly all camcorders have a built-in clock that will allow you to record the date and/or time on your recordings. This function is particularly useful for those who quickly forget the year in which they went to a particular place on holiday. It can also be helpful for reminding you which tape was recorded before another. For those with young children, it gives you an instant way of working out how old a child was in a particular shot.

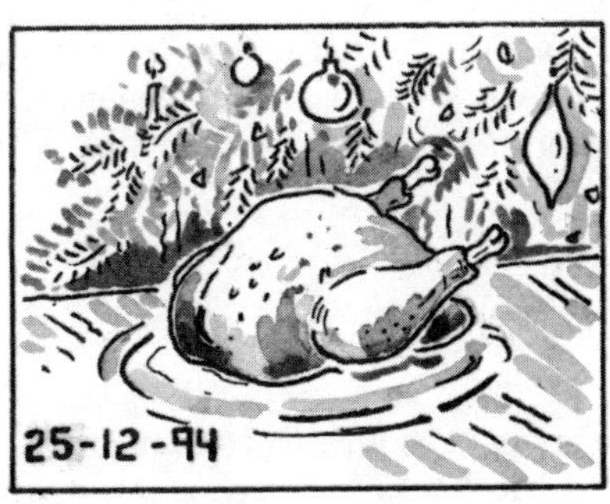

The date can be superimposed on the picture when required

The built-in clock is powered from a separate power supply from the rest of the camera so that it does not have to be reset every time you remove the nicad battery for recharging. Normally you have to insert a small lithium button cell into a sealed compartment in the camcorder to give the necessary power. This cell is normally sold as part of the camcorder package. It should happily last for a couple of years without need of replacement.

By pressing the date/time button when the camera is on, the date and time is displayed in the viewfinder. If you then go into record mode, the date is then recorded in the corner of the picture. However, you should take care not to leave the date on permanently when recording as this can look very irritating on screen. It is a very easy mistake to make, as the date/time can get lost among the other (non-recorded) information that is displayed in the viewfinder. A good habit to get into is to record the date at the beginning of each day's shooting for just a couple of seconds and then turn it off. It is good idea to make this recording when the shot you are recording is unimportant. When it comes to editing your tape you will probably not want to include the date or time, especially if you are going to change the order of events to help structure the story better.

Flying erase head

A flying erase head's job is to wipe any previous recording on a tape just before a new recording is made. It is mounted on the rotating recording drum inside the camera, so that it removes just those video tracks that need to be wiped to make room for the new material. In the earlier days of camcorders, not all camcorders had this facility, and the fixed erase heads produced a distorted image lasting around a second before the picture settled down. Nowadays, the flying erase head is almost universal. It has particular benefit when you want to go back over a recording and add a new section over what has been already recorded.

Filter ring

On the front of the zoom of almost every camcorder, there is a threaded ring to allow you to add filters, or converters, to the lens. Filters can be used for all manner of creative and corrective effects – softening the image for a romantic look, turning lights into stars, cutting out reflections in water and so on. Converters allow you to change the magnification of your lens – a wideangle converter allows you to get more into your picture without moving back, and a telephoto converter magnifies the image. Unfortunately a wide variety of different filter ring diameters are used – so you must ensure that you buy add-on accessories with the correct-sized thread. Some camcorders use obscure sizes that it might be difficult to find accessories for. The most common filter diameters around are 37mm, 46mm, 49mm, 52mm and 55mm. Awkward sizes can usually be accommodated by buying a step-up or step-down ring that converts, say, a 34mm thread into a 37mm one.

AV out sockets

To enable a camcorder to be able to send its picture and sound signals to a television or VCR, it has to have certain output sockets. Unfortunately, these sockets are not standardised. Some sockets have a multi-pin socket that allows picture and sound to be sent along one wire.

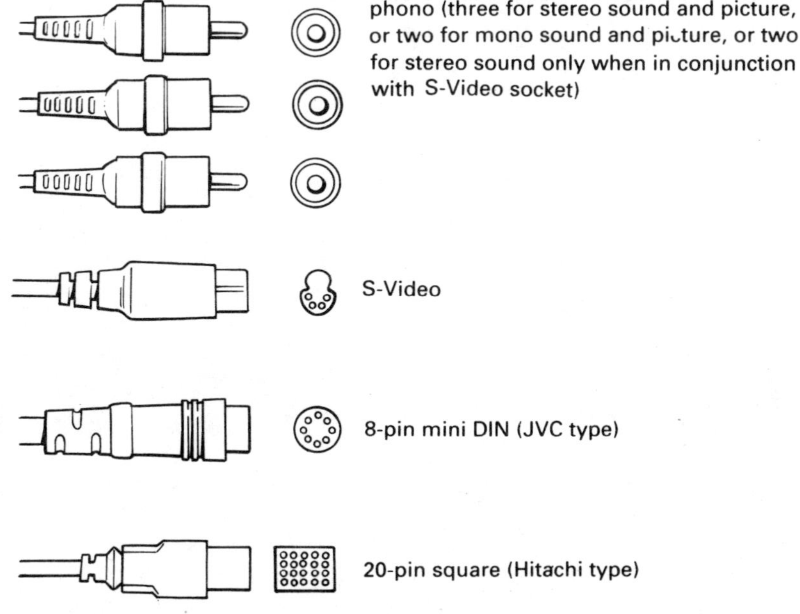

Plugs and sockets used as AV outputs on different camcorders

Others split audio and video signals between different sockets – so you could have two sockets for stereo sound (one for mono) and another for the picture signal. Here are some of the sockets most commonly used on today's camcorders:

Phono sockets – Most commonly used sockets, found on Panasonic, Sony and Canon camcorders. One socket for low-band video (usually coloured yellow). Two sockets for stereo audio coloured red (right channel) and white (left channel) – only one socket used on monaural camcorders.

S-Video socket – Used on most high-band format camcorders (S-VHS, S-VHS-C and Hi8). This special 4-pin plug sends the video signal split into luminance and chrominance to give the maximum possible picture quality from these high-quality camcorders. Audio signal sent via separate sockets (usually one or two phonos)

8-pin mini DIN – Multi-function socket found on JVC camcorders that can send both audio and video signals along one cable.

Hitachi multipin – A 20-pin multi-function socket found on Hitachi camcorders. Outputs not only audio and video signals, but can also output an S-Video signal for high-band models.

4
OPTIONAL CAMCORDER FEATURES

In the last chapter we looked at the features that are found on nearly all camcorders. In this chapter, we turn our attention to the features that are less than universal. Some are found on a good proportion of models, others are specific to just one or two camcorders currently on the market. It is these features that really sort out the men from the boys. The simplest camcorders may have none of the features mentioned in this chapter. The most sophisticated may have most of them.

As pointed out before, some of the features described below are no more than cheap gimmicks – others are essential tools for creative video-making. But as with any purchase, especially if you do not have an unlimited budget, you will undoubtedly find that it will be difficult to find all the features you fancy for the money you want to spend. You will therefore have to make a compromise, a trade-off between different functions. To do this you will need to know the true potential of each of these functions. In this chapter, therefore, we not only explain what each of these buttons does, but how it can help your movie-making.

Manual focus

Autofocus systems can be fooled in a wide variety of situations – by reflections when shooting through glass, by some unimportant object in the foreground of the picture, or even because the subject is made up of a uniform colour without any contrast. More importantly, there may be times when you want one part of the picture deliberately kept out of focus. Or, you may want to shift from having the background in focus to having the foreground in focus while you are recording – to change the emphasis in your shot.

Fortunately, not many manufacturers have been stupid enough to launch camcorders without manual focus controls. However, many have made it a secondary consideration and as a result the controls that make manual focus adjustments can be awkward to use. The best type of focus control is a ring on the lens itself. This gives you the best way of focusing quickly and precisely. The second best form of control is by means of a dial ring, that controls the lens by means of a motor. As you are turning a dial, you can still move focus reasonably quickly, but also make small adjustments when you are close to the correct focus. The most awkward form of manual focus adjustment is by means of a pair of buttons – one for making the focused distance shorter, the other for making it longer. If you have to settle for this last type of control, at least ensure that the buttons fall readily to hand.

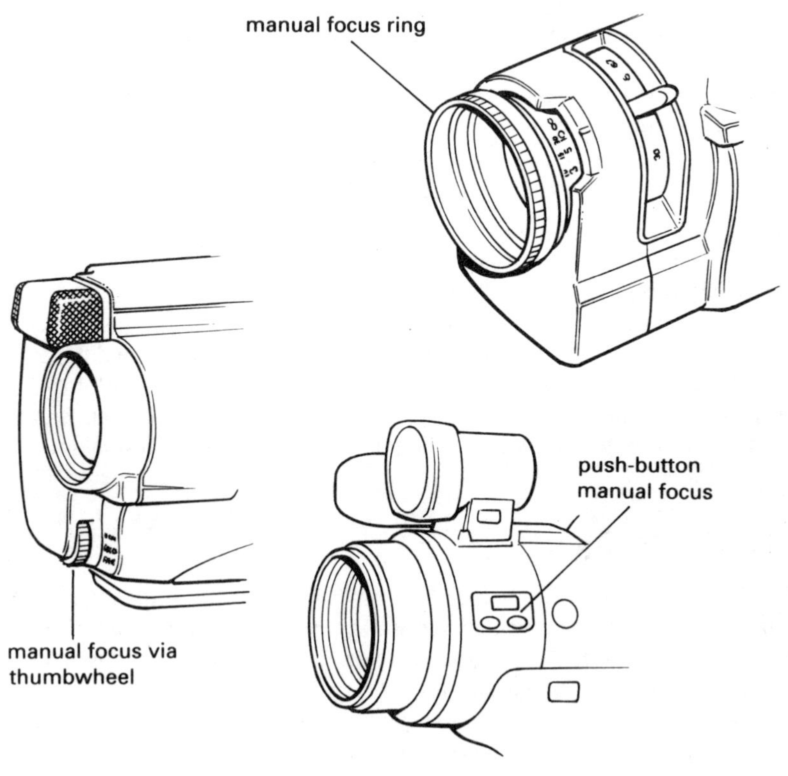

Manual white balance

All camcorders have an automatic white balance setting – making sure that you get natural-looking colours under a wide variety of lighting conditions. Although the white balance system on camcorders is generally good at getting the right results, there are times when you want it to get it wrong deliberately. A common case might be when you are videoing a sunset. You want a full range of reds and oranges in the picture – and you don't want the AWB system to filter these out.

A manual white balance control will give you a selection of preset white balance settings – such as daylight, tungsten (normal bulb lighting) and fluorescent (strip lighting). Not only can you use the daylight setting to ensure that the camcorder does not overcompensate for sunsets, as in the above example, but you can use these presets for special effect. For instance, if you use the tungsten setting in daylight, you can, with care, get a blue picture that looks similar to moonlight. Alternatively, in candlelight you might use a daylight filter to boost the orange glow of the candles.

Some camcorders give you true manual white balance control – you set the white balance for a particular light source. This is done by holding something white in front of the camera under the light you want to set the system to, and then press the white balance lock. This can be particularly useful when your subject is lit by a variety of different light sources – such as bulb lighting mixed with daylight. You can then choose which light source you want to predominate. We will be looking at the whole subject of white balance in more detail in a later chapter.

Manual zoom

Motorised zooms have their uses, but they are limited by only have one or, at most, two speeds of operation. Sometimes, the most effective zoom during recording is the creep zoom – where the lens zooms in (or out) on the subject so slowly that it is almost imperceptible to the viewer. This has to be done with a manual zoom and a very steady hand. Conversely, a sudden 'crash' zoom can occasionally be used for dramatic effect.

It has to be said that the zoom lens should mainly be used for framing up a shot before you start recording. As such, using the motorised zoom is an unnecessary waste of valuable battery power. Your batteries need recharging soon enough without wasting energy. In certain situations too, you need to move from telephoto to wideangle, or vice versa, as quickly as possible before the subject disappears altogether. Waiting for the motorised control to get to the right focal length could mean missing the shot – a quick twist of the manual zoom lever, however, and you are ready for action.

Manual iris

In the last chapter we saw that the backlight control provides an easy answer to avoiding silhouettes when shooting against a window. However, this type of button has severe limitations. It can only make the one exposure correction – even with the window, it might still not let in enough light. A manual iris gives you complete control, not just over this situation, but whenever the auto exposure system is likely to have problems. For example, if you are shooting a white cat against a white wall, the automatic system will tend to turn both a greyish colour. Conversely, if you are shooting a very dark scene, such as a cityscape, at night, the camera may try and boost exposure so that none of the scene remains black.

We will also see in later chapters that by opening and closing the iris you can control how much of the picture is in focus. This is called depth of field, and means that you can throw a distracting background out of focus, or ensure that everything from your feet to the horizon is completely sharp. A manual iris gives you complete control over this phenomenon.

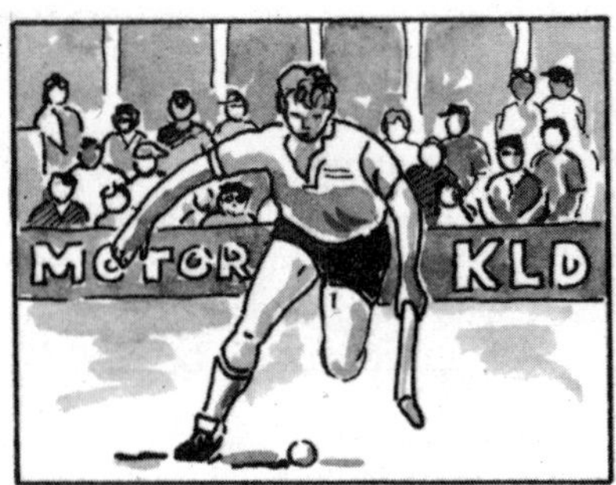

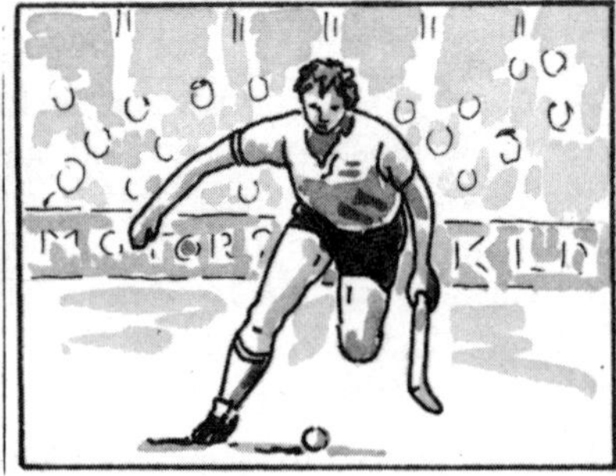

A manual iris can help you make a distracting background (left) look like an out-of-focus blur (right)

Manual gain

As described in the last chapter, one of the ways that a camcorder can increase exposure is by increasing the gain – amplifying the video signal received by the CCD imaging chip before it is recorded. Some camcorders allow you to have manual control of this feature – so that you can decide whether or not you need to use it in low light. Sometimes this is useful, as you may prefer a slightly darker picture to one which is lighter, but also grainier.

Programmed exposure

Programmed exposure is a common facility on stills cameras, but has to be treated with a certain amount of caution on camcorders. In programmed exposure, the camcorder takes complete control of shutter speed, iris, and gain controls. Typically a range of programs is available on a particular camcorder – each catering for a particular situation. For instance, there might be a portrait mode, that always tries to provide the widest aperture possible, so that only the subject is in focus, and distracting backgrounds remain blurred. Another common program mode is for sport – where the shutter speed is set slightly faster than normal (say between 1/100 and 1/500sec) to eliminate some of the blur in the picture without getting playback that is too jerky (which comes from using the fastest shutter speeds, designed for frame-by-frame analysis, for normal speed playback). More specialist programs can be found for tackling such varied subjects as golf, skiing, night scenes or spotlighting.

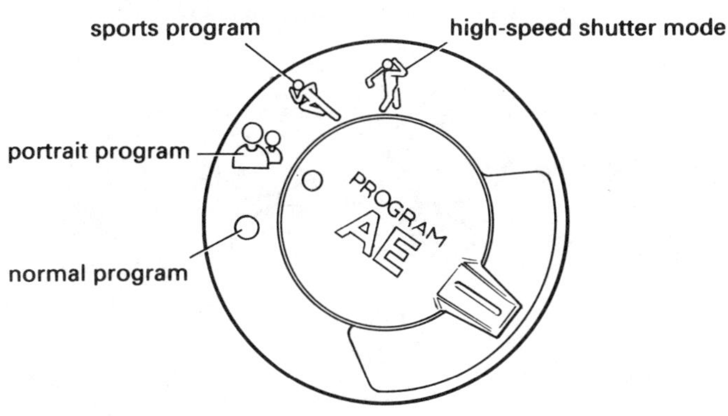

A typical program exposure dial

The trouble with all of these program exposures is that they are excellent in some situations but rotten in others. If you are going to learn when and why they are going to work well, you might as well have full manual control over exposure. For example, a portrait feature is fine on many occasions, as you want the background out of focus. However, for some portraits you might need to see what is going on in the background. Also, you should realise that in bright light the large aperture is made possible by using a high shutter speed. Therefore if there is anything moving in the frame, the movement could appear jerky on playback. In short, programmed exposure settings are meant to help those who do not want to bother learning about manual settings – however, using these settings without any such knowledge could lead you to worse-looking footage than if you had used no programmed (or manual) settings at all!

Interchangeable lenses

Although commonplace in the world of professional video, only one or two consumer camcorders have truly interchangeable lenses. The best known is made by Canon. The advantage is that you do not have to commit yourself to a built-in zoom. A choice of zooms is possible, from say one with a 15x zoom ratio, with a powerful telephoto setting, to one that is designed for wideangle work. Fixed focal lengths are also possible – such as lenses particularly designed for close-up work, or for astronomy. You can also, by means of an adaptor, add lenses designed for SLR stills cameras. However, as a CCD chip is much smaller than a film negative, even a wideangle lens becomes a telephoto lens when bolted on a camcorder. Interchangeable lenses are very expensive.

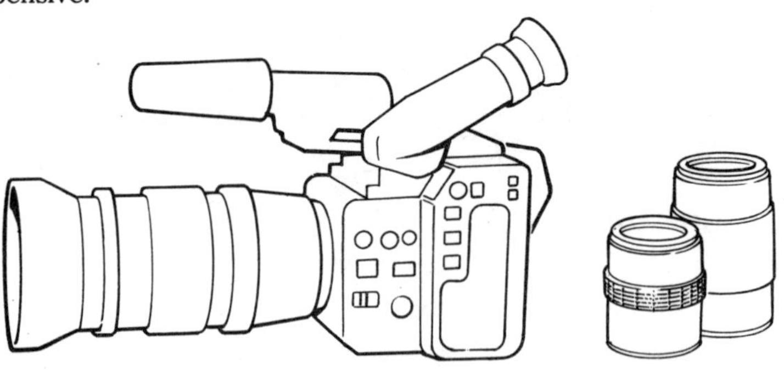

Canon's EX-2 Hi camcorder with interchangeable lenses

A couple of manufacturers have used a semi-interchangeable lens system to add a choice of lens to their camcorders, without the cost of the lenses being too high. Here, just the front elements of the lens are changed, to give a range of different possible zooms – but all with the same zoom ratio. A good halfway house, that is particularly useful for ensuring that you have a good wideangle setting at your disposal.

Macro

A manual macro control is a lens setting used for recording small objects at very close range. On some camcorders, in fact, the subject can be just a few millimetres from the lens' surface – giving massive blow-ups when viewed on screen. The macro control is usually found on a manual zoom ring. By releasing a lock at the wideangle end of the zoom lever's range, the macro mode is engaged. The picture is then focused by making small adjustments with the zoom lever. On camcorders without a zoom lever, the macro mode is normally engaged automatically when needed (but normally only if you are at the wideangle end of the zoom range).

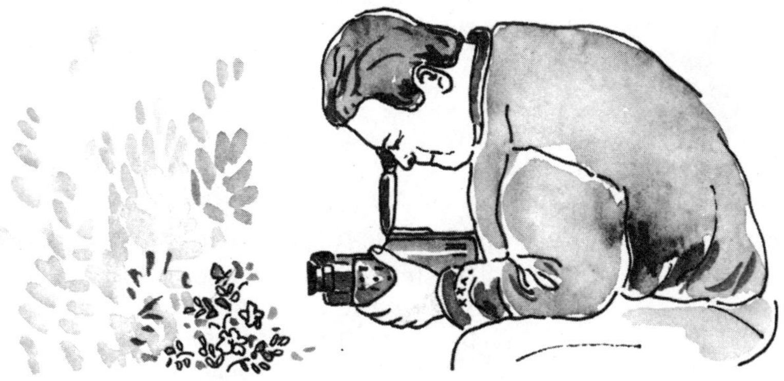

Most camcorders allow you to focus on objects just a few millimetres away when using the wideangle setting

Edit button

The edit button is a small switch that is generally found near the transport controls of the camcorder. It is only used when using the camcorder as the playback deck while editing, or copying, a tape. It turns off the picture enhancement circuitry normally used during playback on a camcorder (or a VCR). You do not want this enhancement to be recorded, as it will be added when your copy, or edited copy, is replayed through the VCR.

Audio dub

The ability to audio dub is one of the main advantages of the VHS formats (VHS, VHS-C, S-VHS and S-VHS-C). Basically, an audio dub allows you to add a new soundtrack to your tape, without having to re-record over the picture. It is made possible because all four of the VHS formats have a mono linear soundtrack, which is recorded along the side of the tape – in the same way as music is recorded on an audio cassette. Because it is completely separate from the video tracks, it can be re-recorded at a later date without losing the picture.

Audio dubs are not possible on 8mm and Hi8 camcorders, as the soundtrack, whether it be mono or stereo, is recorded in the same place as the picture. To change the sound you also have to change the picture. The simplest way to add music or commentary, therefore, is to add the new sound as you make a copy of the video – as part of an editing process. Alternatively, the 8mm or Hi8 user makes a copy of the recording on VHS or S-VHS, and then uses the audio dub facility on the VCR.

The advantage of the VHS formats, therefore, is that you can change the soundtracks without having to copy the picture. As any copying of the video signal means a loss of quality, this can be very useful.

An added bonus comes to those using one of the VHS formats, if they also have stereo sound. The stereo is in fact a second soundtrack – you also have the same sound recorded on the mono linear track. Although you cannot directly replace the stereo soundtracks (for the same reason as you can't directly replace the soundtracks on 8mm and Hi8 recordings) you can still replace the mono soundtrack. Therefore, you are not losing the original sound – this stays on the stereo tracks. Adding music, commentary or whatever to the mono linear track becomes a bonus – not a complete replacement. On stereo VHS family camcorders you also get a switch that allows you to choose which of the two soundtracks you want to listen to on playback – you can choose the mono linear track on its own (often labelled 'Normal'), the stereo tracks on their own (often labelled 'HiFi') or play both of them simultaneously (often labelled 'Mix').

Of course, not all VHS-type camcorders have the audio dub feature. Although just because it is not a feature of the camcorder itself does not mean that audio dubs are not possible. Some JVC camcorders, for instance, only give you audio dub when you buy an optional remote control unit. Also worth remembering is that although you can add an audio-dubbed soundtrack using the camcorder's built-in mic, it is much better to be able to plug a tape recorder, or special speech microphone, directly into the camcorder's mic socket – if it has one, that is!

There is an exception to the rule that says that all 8mm and Hi8 camcorders do not have audio dub – and that is the one or two Hi8 camcorders that have a second PCM soundtrack. This digital stereo track is recorded separately from the normal stereo soundtrack, and because of this can be replaced without affecting the picture.

Insert edit

An insert edit facility often goes hand in hand with an audio dub control. Again, the key to its operation is the mono linear soundtrack on the four VHS formats. What it does is allow you to re-record the picture on a tape without affecting the original sound (as recorded on the mono linear soundtrack). A new soundtrack is recorded if the camcorder is a stereo one, but on the stereo soundtrack alone (this can be mixed, or cut out completely, using the Normal/HiFi/Mix button as with audio dubs).

An insert edit is an incredibly useful device for covering up bad shots, or boring bits, in a video that you have just made. You could record over them completely, but the break in the soundtrack could give the game away. Perhaps you are recording a wedding, but as the bride and groom come out of the church your view is obscured for a few seconds by a bystander. If you re-record the shot completely, there will be a jump in the sound, as the bells are ringing. A better solution is to leave the mistake till later, when you can cunningly insert edit an innocuous shot of the church or the car to hide the offending shot of the bystander.

Again, not all VHS family camcorders have this facility – and, again, some camcorders only allow you to access this facility using an optional remote control unit.

Confusingly, some 8mm and Hi8 camcorders claim that they have an insert edit facility. This is not the case. What they are referring to is an ability to replace sound and picture over a section in the middle of a tape you have recorded. This is not the same thing at all.

Mic socket

As explained in the last chapter, the built-in microphone on any camcorder is a jack-of-all-trades – and, as the adage goes, it is a master of none. Sometimes you will need the help of a specialist microphone to be able to get the sound you want and this is where the provision of a mic socket is vital.

Microphones are available that are customised for speech, or for picking up faint sounds a hundred yards away or more. If you have a stereo microphone, you might want to enhance the three-dimensional stereo effect by using two mics placed a fair distance apart. We will be looking at add-on microphones, and what they can do, in a later chapter.

A mic socket can also be useful on other occasions – for audio dubs, for instance, as explained above. Also, you can mix music and speech through a special audio mixer, or mixing mic, as you record your picture, using the mic input.

Narration mic

A second built-in microphone found on some top-end camcorders, carefully positioned so that you can add your own commentary as you shoot.

Headphone socket

The human brain is an incredible instrument. You can be sitting in a noisy bar, but you hear what your friend is saying perfectly because your brain filters out the unnecessary sound. A microphone is not as sophisticated and cannot hope to know which are the important sounds and which are not, and so the camcorder records all that the mic picks up. The only way you can be sure that you are recording the sound you want to record is to monitor it carefully using an earpiece or pair of headphones. To be able to do this you will need a headphone socket. Even the best video can be ruined by a muffled soundtrack.

Monitor speaker

Only found on semi-professional on-the-shoulder models, a monitor speaker allows you to listen to the sound you have recorded (though not in stereo) as you playback your footage. The small speaker is positioned by your right ear as you hold the camcorder.

Timecode generator

In the world of professional video, a timecode generator is an essential feature on any camcorder. For the rest of us, it is a useful feature to consider if you are likely to become heavily involved in editing your videos. What the timecode generator does is to record a unique number on every single frame of your video. These numbers are recorded separately from the picture, so do not have to be shown on screen when you replay your video.

Using a suitable edit controller, each frame can then be found and identified precisely when editing. So when you copy a new sequence you can guarantee that it will start and stop with the exact point you intended. Without timecode, precision editing is a hit-and-miss affair – relying either on the camcorder's less-than-accurate tape counter or on the manual dexterity of the operator.

There are two different forms of timecode commonly found on today's camcorders – and unfortunately they are incompatible with each other. VITC (pronounced *vitsee* and standing for Vertical Interval Time Code) is found on camcorders using the four VHS formats. Even if these camcorders do not have a VITC generator built-in, many offer the possibility of using one as an optional extra – even for some budget-priced machines. VITC must be added at the time of initial recording – it can not be added subsequently without making a second-generation copy of the footage.

More recently, Sony have introduced RCTC (which stands for Rewritable Consumer Time Code, and pronounced by some as *arctic*). This is used on a small number of Sony's and Canon's top-of-the-range Hi8 camcorders. Its advantage over VITC is that it can be added to the recording at a later date if necessary and so can be superimposed on all your old 8mm and Hi8 recordings.

Edit terminal

An edit terminal is becoming an essential feature to have if you want to take full advantage of the increasing number of automatic edit controllers that are coming on the market. It is basically a remote control socket that allows another machine to take control over various of the camcorder's functions. Equally as important is the edit terminal's ability to send, as well as to receive information. In order for an automatic edit controller to function, it must be able to read the camcorder's tape counter, and it must be able to take control of the camcorder's tape transport mechanism.

There are two types of edit terminal commonly found on today's camcorders. The Control-L socket is found on nearly all camcorders made by Sony, and on some others – such as some models made by Canon and Sanyo. The RMC 5-pin socket is found on many camcorders made by Panasonic.

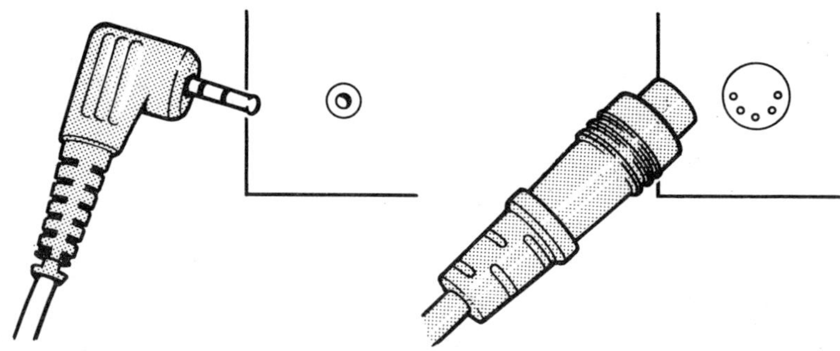

Control-L plug and socket   Panasonic 5-pin edit connector and socket

Synchro edit socket

Although too few camcorders have a proper edit terminal, there are many that have a so-called synchro-edit socket. Although this is of some use for editing, it is far more limited in its application. Alternatively known as remote pause, synchro editing basically simplifies the number of buttons you have to press when editing from a camcorder to a VCR which has a compatible synchro edit control socket. You press the play button on your camcorder, and the VCR goes from record pause into record mode. The biggest headache of this system is ensuring that your camcorder and VCR operate compatible synchro-edit systems – the best way to help ensure this is to have a camcorder and VCR made by the same manufacturer (although this is not foolproof). You are better off trying to buy a camcorder with a proper edit terminal.

Title generator

There are two different systems for adding titles that are commonly found on camcorders. One works a bit like a word processor – allowing you to string individual letters together on screen to make up you title – this is known as a title generator. The second system works by you writing or drawing your titles by hand, and then storing these electronically in the camcorder for use later – this is known as a digital superimposer (see below).

The title generators that are built into camcorders are simple affairs. Generally they will only allow you to have two lines on the screen at once – although some do allow you to store two such titles or 'pages'. Some will allow you to choose the colour of these titles from a limited palette. Others will allow simple 'scrolls' where the title will move, say, up from the bottom of the picture to the centre of the screen.

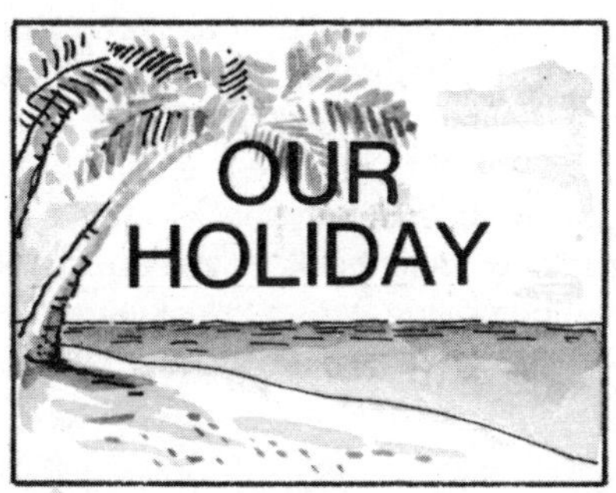

Character generators allow you to type in your own titles for your movies

Despite being limited in its scope, the beauty of a built-in titler is that you can punch in your titles before you start shooting – storing them for when you need them. They can be called back from the camcorder's memory when they are actually needed – and are only recorded on tape when you go into record mode when they are displayed in the viewfinder.

If your camcorder does not have such a titler built in, you may find that one is available as an optional extra. Alternatively, titles can be added during the editing process using a wide variety of stand-alone title generators – or by using a specially adapted computer.

Digital title superimposer

A digital superimposer is primarily intended for adding titles to your videos, but can also be used for a variety of special effects. Basically, by pressing the memory button of the superimposer the camcorder stores anything that is in front of the camera – not as a real picture, but as a high-contrast black-and-white outline. Once stored, the camcorder can then change the colour of this blocky image. Like a title generator, the titles can be stored until you want them and are only recorded when you go into record mode with the stored frame displayed in the viewfinder. Some models allow you to store two such frames, and may allow you to choose from a palette of up to eight colours. As with a built-in titler, a limited number of scrolls may also be at your disposal.

The most common way to use the digital superimposer is to write your title on a piece of paper and then memorise this while shooting it in macro mode. Where the facility comes into its own is when you want to record a piece of prepared artwork – such as a company logo, or a drawing. Commonly, the stored frame can be reversed – so instead of seeing the video picture behind the coloured title, the title itself becomes a keyhole through which to see the live action, while the rest of the frame is in solid colour. Interesting special effects are therefore possible – such as being able to make it look as you are shooting through a pair of binoculars – by memorising two interlocking circles, and then reversing. A more adventurous application might be to superimpose

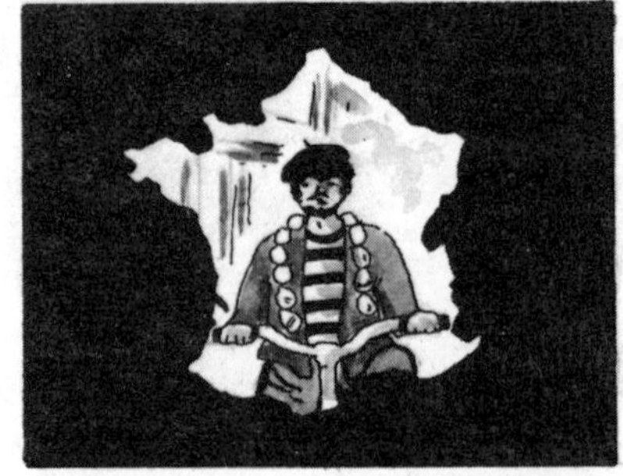

Here a digital superimposer has been used to make a window the shape of the map of France

the outline of a map of the USA and shoot through this while videoing the introduction to your video of Disneyland.

Digital zoom

A camcorder that boldly boasts that it has a 100x zoom is using electronic trickery to make this possible. In truth, the telephoto end of the lens is not 100x more powerful than the wideangle end. Instead, the camcorder probably has a 10x optical zoom. However, when it reaches the end of its true telephoto ability, it is possible to switch over to the electronic zoom. With this the camcorder successively enlarges smaller and smaller parts of the centre of the image.

As with a photographic enlargement, the bigger the blow-up the more the grain shows – or, in the case of the camcorder, the pixels show. The result is coarse image – where you can see the individual dots that make up the picture. The amount of detail in the picture is greatly reduced – making the facility practically useless. As if this was not bad enough, even the tiniest movement of the camcorder at this degree of magnification shows up as a severely shaky picture. Even the most solid tripod cannot hope to rid you completely of this involuntary movement.

The best effect it can be used for is for zooming out from a faraway object, such as the moon. As long as the picture is not held for long on the mega-close-up, you get the impression of moving away from the moon in warp drive.

Black and white

A simple digital effect which allows you to record the picture in black and white. Other camcorders have a similar sepia effect mode, where the picture is recorded in orangy-brown tones – similar to Victorian photographic prints.

Paint

A digital effect that looks impressive the first time you see it, but has very limited use in practice. What the camcorder does is to limit the number of colours that are recorded by breaking the picture up into blocks of tone. The effect can often be seen in pop videos. Also known as solarisation on some camcorders.

Mosaic

Another interesting-looking digital effect that has little practical use. Here the camcorder effectively enlarges the size of the pixels, giving a picture that is made up of squares of uniform colour.

Mirror

A digital effect that is designed for fun rather than for serious use. Here the right-hand side of the picture becomes an exact mirror image of the left-hand side – with hilarious consequences when videoing people's faces.

Freeze frame

Digital effect where instead of taking moving pictures, the camcorder records a brief still shot lasting a few seconds. The sound is generally recorded as normal. Useful for freezing the action for the end of a sequence, perhaps – or to make a gallery-type sequence of shots as a special effect. It can also be used for recording artwork and pictures without worrying about camera shake. Also known as digital snapshot or digital frame store.

Dissolve

One of the most useful digital effects that you can have. Also known as digital mix, a dissolve allows you to move from one scene to the next by one picture being gradually replaced by the other. For example, if you were moving from a shot of a building to a close-up of a face, you would only see the face very faintly to begin with, but this would become more and more identifiable as the shot of the building faded away completely.

The camcorder does this by freeze framing the first shot (see above) and gradually strengthening the video signal of the second, live, shot.

Timebase corrector

A timebase corrector is a standard stand-alone accessory in any professional edit suite, but is also found built into a few top-end consumer camcorders and VCRs.

As well as recording sound and pictures on video tape, every camcorder also records a synchronisation (sync) pulse. For every picture frame there is one of these pulses. These are used when playing back a tape, to ensure that the video head accurately recognises where each video track begins – it is like the electronic equivalent of sprocket holes on movie film. This system works very well, until the tape deteriorates in some way – most commonly when the tape is a copy of a copy of a copy. This results in these sync pulses becoming weak. Then when the tape is replayed the synchronisation of video head and tape is out. The result is an inferior picture – the most notable feature of which is that vertical lines which should be straight, begin to become ragged or even diagonal. The job of a timebase corrector is to replace these sync pulses digitally so that deteriorated tapes can be played without significant quality loss.

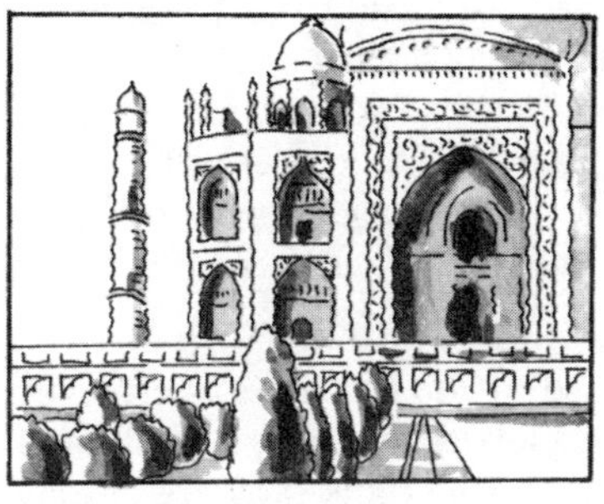

without TBC

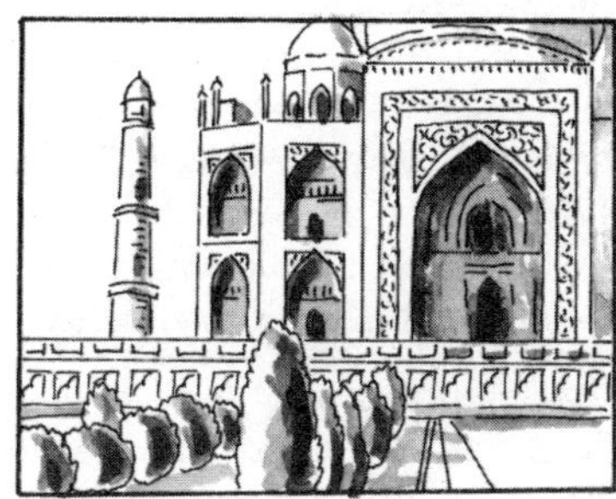

with TBC

Timebase corrector

Slow shutter effect

All camcorders record 50 pictures a second to give the illusion of movement (60 per second on the American NTSC system). Usually, therefore, the effective shutter speed of the camcorder is 1/50th of a second (1/60sec in US). As we saw in the previous chapter, most camcorders can use a faster shutter speed if so required. However, a few camcorders also have the ability to give the impression of using a slower shutter speed. This digital effect is achieved by recording several successive frames on top of each other. When used in lowlight, the effect gives curious looking results when played back. Moving subjects look blurred – ideal for videoing car lights on a busy road at night, or for interesting footage on the dance floor. One of the more useful digital special effects that are available.

Tracer

A digital effect that has a similar effect to a slow shutter speed. The tracer effect, however, overlays each frame with ghost images of the frames that have gone immediately before. This gives an impressive streaking effect on any moving subject in the frame.

Quasi high-band

In normal circumstances, it is not even possible to play a high-band (Hi8, S-VHS, S-VHS-C) format tape on a low-band (8mm, VHS, VHS-C) camcorder – let alone get the superior picture quality that these tapes hold. However, manufacturers are increasingly adding a confusing feature that gets round this problem. On an 8mm camcorder, for instance, the quasi-Hi8 system will allow you to play a Hi8 tape. You will not get Hi8 quality – but you do get watchable picture and sound, which is similar in quality to an 8mm recording. Similarly, a quasi-S-VHS facility is increasingly found on VHS video recorders.

This feature is only going to be of any use to you if you are likely to be given high-band recordings to view from a fellow camcorder enthusiast. Otherwise it can cause confusion to buyers who think they are getting true high-band recording. Even more confusingly, some manufacturers refer to this facility as Hi8 (or S-VHS) playback.

Twin lens

Having two lenses on a camcorder was an innovation introduced by Sharp in 1991. It is a clever idea as it allows the camera to have not only a normal zoom lens, but a proper wideangle lens as well – without having to increase the length of the body. Each lens has its own imaging chip – and you can switch between the two at the touch of a button. Because of this, it also comes in useful for making quick shot transitions

– cutting, say, from a full wideangle shot of a landscape to a close-up telephoto shot of a person in the distance. To avoid the ugly jumps that such a transition can cause, the Sharp camcorder also allows you to fade or wipe between these two shots.

Clip-on light

Although camcorders can record almost in total darkness, any improvement in lighting will dramatically improve the picture quality. Increasingly, manufacturers are providing miniature clip-on lights with their camcorders. Some have even have gone as far as building a light into the camcorder chassis itself.

These lights are very low-powered, typically only giving out five watts of light. However, even such a small lamp will do wonders in poor lighting – but only if the subject is just a few feet away. The drawback is that they run directly from the camcorder's rechargeable battery – normally halving the amount of recording time you can get before recharging or using a second battery. A useful device for emergencies.

Flexigrip

The viewfinder on most camcorders is tiltable to some degree, so that it can be raised to the most comfortable angle for shooting. Some

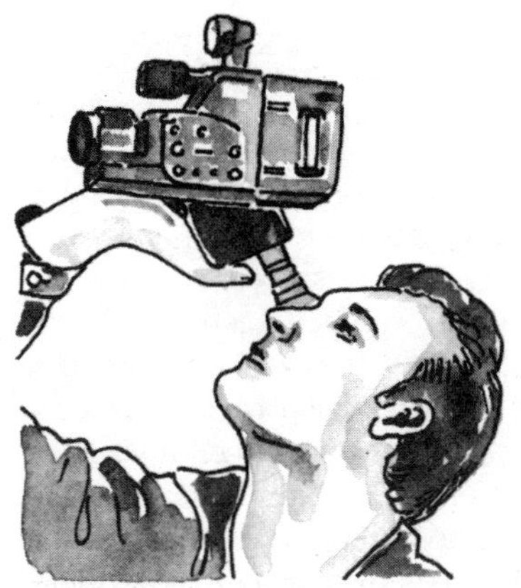

using camcorder above the head

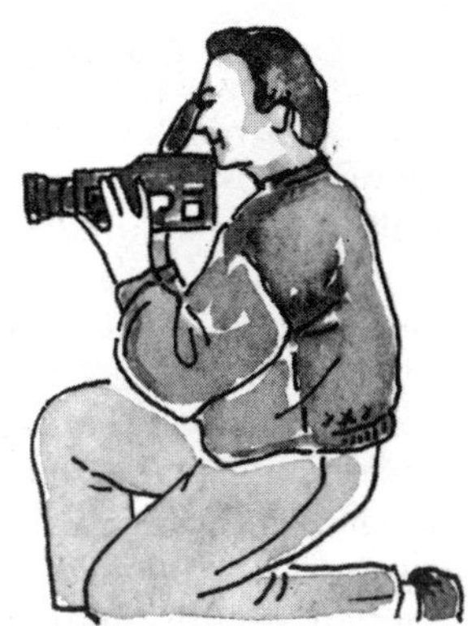

shooting while kneeling

Uses for a flexigrip viewfinder

models, however, have a viewfinder that can be rotated by up to 180°. With this much rotation, it makes the camera usable even when it is lying on the ground, or raised above the head – as the viewfinder can be positioned in easy view of the user.

Handgrip

Normally a camcorder is supported by means of an adjustable strap on the right-hand side of the camera. If the camera is large and heavy, extra support is given by a shoulder rest. A few camcorders offer a second way of holding the camera – by means of a handgrip, positioned underneath. Some people find this sort of handle more comfortable to use and it has the advantage that it can be held by the left-hand, a token concession to southpaws.

Age insert

As well as having the usual date and time function, some camcorders also allow you to program in your child's birthday. In this way, when required, you can record your child's age (instead of the date) at the beginning of a scene.

Splashproof casing

It almost goes without saying that every camcorder is packed full of delicate electronics which can be quickly destroyed by water, sand or dust getting into the casing. For this reason, a number of manufacturers have introduced models which are specifically designed to withstand particularly adverse conditions – such as ski slopes, beaches or boats. Most of these models are merely splashproof – they will not withstand a complete dunking. The extra plastic casing, which is often brightly coloured, makes these models bigger and heavier than their less-protected counterparts. Worth considering, perhaps, if you are going to use your camcorder in such conditions most of the time. But remember, they do not give 100% protection. It is also worth pointing out that splashproof casings, and even completely waterproof housings for underwater use, are available for many models as optional accessories.

Inner focus lens

An inner focus lens is one that has no external moving parts – for example, the front ring of the lens does not rotate as you focus. The downside is that these lenses do not usually have a manual zoom ring or a true manual focus ring. Some internal focus lenses make up for this last omission by retaining a front focusing ring – that can focus via the motor inside the lens.

Home station

A home station is a system for simplifying the number of connections necessary for charging the battery and/or linking your camcorder up to the television. Once the home station is connected to the mains and to your TV inputs, a task that only has to be performed once, the camcorder can be simply slotted on the station, forming a direct link to the relevant contacts on the base of the camcorder.

Animation

Strictly speaking, animation is not possible with video cameras. Professional cartoons are made using film, with the picture changing 24 times a second to give the impression of movement. Unfortunately, it is not technically possible to control a camcorder precisely enough to get such short picture bursts. However, it is possible with some camcorders to do rough animation by recording slightly longer bursts in succession – around eight different frames per second. You could produce an animated film of a 200-piece jigsaw miraculously putting itself together. Using a special animation mode, you would put the pieces of the jigsaw in place one by one – recording each for 1/8sec using this special recording mode. The finished film would therefore last just 25 seconds in total.

Time lapse

A time lapse facility works in a very similar way to the animation mode. Instead of having to start each short shot manually, however, each brief recording is triggered automatically after a preset period measured by the camcorder. You could use this facility to speed up the recording

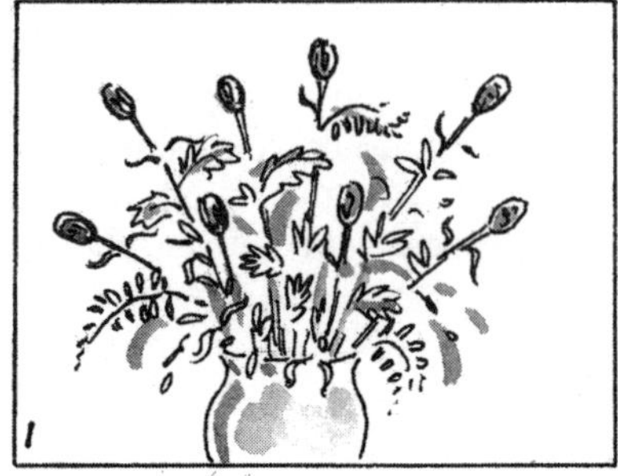
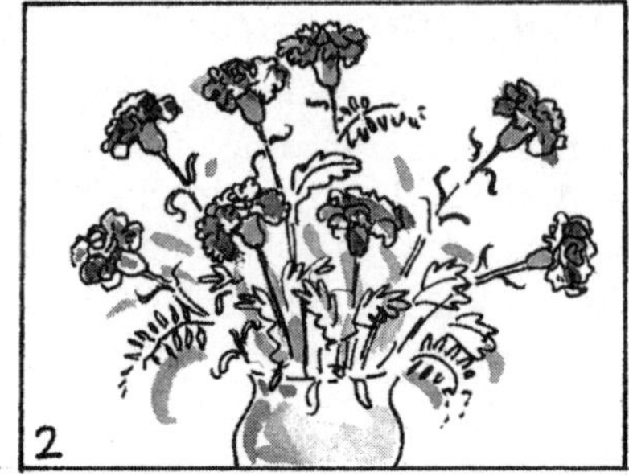
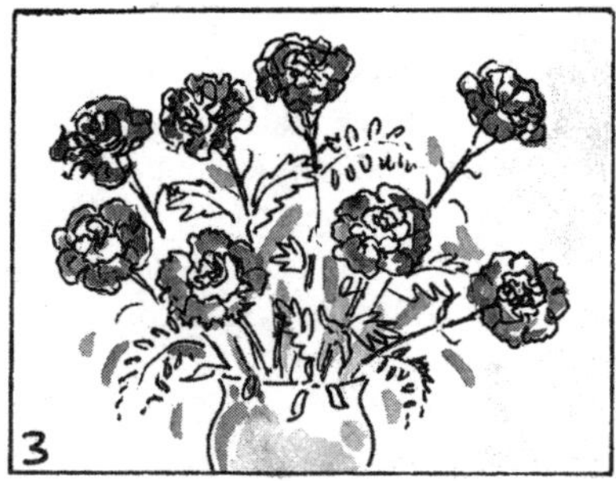

Using an intervalometer condenses an event that takes place over several hours – such as flowers opening – into just a few seconds

of a flower opening, for instance. By making a recording lasting 1/8sec every minute, an event that normally took place over two hours could be condensed into just 15 seconds of video. As the individual shots are much longer than with normal video, the results can look jerky, as with video animation.

Selftimer

A selftimer allows you time to move from behind the camera to in front of the camera before recording begins. Generally there is a delayed start of about 10 seconds. Not such a useful feature as on a stills camera, as you can start recording anyway and then join the rest of the group – at the very worst all you will have is a short bit of wasted video as you move into place. Remote controls are becoming an increasingly common accessory that comes with the camcorder – allowing you to start the camcorder from a distance – so a selftimer is becoming even more redundant.

Remote control

A remote control has two principal functions on a camcorder. It can be used to start recording from a distance when shooting – when you want to be in the picture yourself, or when you want to video shy wildlife from a close distance. Secondly, it is useful when playing back your tape on a television – allowing you to control the tape without having to get up and fiddle with the transport controls of the camcorder itself. As well as having record and tape transport controls, the remote might also allow you to control the zoom lens.

Some manufacturers also 'hide' less-used buttons on the remote control. JVC, for instance, often put their insert edit and audio dub buttons on the handset.

Other functions that might only be found on the remote include a basic automatic editing control – where with a compatible VCR you can pre-choose segments from your original footage to be automatically copied. A very useful feature.

Image stabilisers

One of the major problems with the trend towards smaller and lighter camcorders, is that while they are easier to carry, they are also harder to hold steady. Although many books will tell you to use a tripod whenever possible when shooting, this cumbersome lump of metal tends to negate the advantages of having a portable video camera. Unfortunately, camera shake makes your footage look amateurish, and is uncomfortable for your viewers to watch. There is little doubt that this problem is the most common fault made by the amateur video maker.

To help avoid camera shake, therefore, some manufacturers have come up with systems which compensate for this involuntary movement. There are two different image stabilisation systems used. The first, and most widespread, is a digital system. By comparing the live picture, with the one that was taken immediately before (which is stored in the digital memory), the system can work out whether the camera has moved and compensate accordingly – by using the part of the image that matches the one that has gone before closest, and enlarging this part digitally to fill the whole frame. The enlarging process does mean a slight decrease in picture quality – but this is much to be preferred over a shaky picture. In any case, the digital image stabilisation system can be turned off if not required – which is necessary as the system can try to compensate for deliberate camera movement.

The second, and more recent, image stabilisation system works using an entirely different principle. Using movement sensors inside the camera, a pair of lens elements in the zoom itself are constantly adjusted to keep the image steady. The advantage over the digital system is that there is no degradation of the picture. The main disadvantages are that the system is more expensive, and makes the camcorder heavier.

I-HQ

Standing for Intelligent High Quality, I-HQ is a picture enhancement system pioneered by Akai. It works by optimising the camcorder's recording system to the grade of tape being used at the time.

Accessory shoe

An accessory shoe is a platform on which add-ons such as microphones, lights and timecode generators can be clamped – so they don't have to be held separately when using the camcorder. The shoe looks similar to a hotshoe on a stills camera, except that there are no electrical contacts for direct connection to the camcorder. At one time, nearly every camcorder had such a shoe. But as the camcorder has become smaller, manufacturers have found it harder to find space to put it – so have tended to leave it off. Although it is very useful, you can buy accessory shoe brackets which will fit to most camcorders by slotting between the battery and the battery contacts on the camera.

Tally light

One of the most important things to pay attention to when using a camcorder is whether you are recording or not. The viewfinder provides a moving picture whether you are in record or pause mode – and normally all you have to guide you as to which of the two your camcorder is in is whether the word 'Record' or 'Rec' appears in the viewfinder or not. As there is often a lot of other information in the viewfinder, it easy to mistake which mode you are in – with dire consequences.

An added indication of whether you are in the record mode is a tally lamp – a small red light at the front of the camera that switches on when you go into record mode. This provides valuable reassurance to the camera operator, but can have its disadvantages. It can give the game away when videoing the more self-conscious members of your family – and can show up as a reflection when shooting through glass, especially at night. A camcorder with such a tally light should also allow you to turn it off completely when required (although a little piece of black insulating tape could cover it).

Built-in monitor

Built-in LCD colour monitors have two purposes. Firstly they can be used as a viewfinder while shooting. As the screens measure three or four inches across you can watch what you are shooting with the camera held away from your face – so you can surreptitiously shoot from the hip, or even hold the camera above your head. However, this style of shooting can take some getting used to, so a more conventional view-finder is also beneficial. The second use is for playing your footage back. Unlike with a conventional EVF, several people can watch at the same time – and there is usually a built-in speaker so you can hear the soundtrack too. This can be very useful if you are using your camcorder miles away from a suitable TV. However, LCD monitors are available as accessories for most camcorders.

5
—— BUYING ADVICE ——

For most people, the most important consideration when buying a camcorder is to get the best value for money. But there is more to achieving this goal than walking into the first shop you see and buying the cheapest model in stock. Even if you have a particular model in mind, getting the best price doesn't automatically equate to the best deal. It might be cheaper at shop A, but does it come with the same accessories and is the warranty the same as at shop B?

The first thing you should do is some research. Does one particular tape format suit your needs best – or is this issue unimportant to you? What features do you feel are essential for the type of videos that you are going to shoot – which features would you like as well, if at all possible? If you have read this book up to this point, you will already be in a position to begin drawing up a shopping list that answers these questions. But your research is still not complete. You could walk into the first shop you came to and find that they had a camcorder in stock that filled all your requirements. However, you would have no idea that there was an alternative model that would suit you as well, have even more useful features, and was even cheaper.

A wise move, therefore, is to buy a specialist camcorder magazine before you go out shopping – such as *Video Camera* in the UK, or *Video Maker* in the US. Not only will such magazines provide you with a list of everything that is currently available, with an overview of their main features, their advertisements will also give you a guide to some of the discounts that are available on certain models. They may also give you news of camcorders that are not yet in the shops, that it might be worth waiting a few weeks for.

Although you can buy your camcorder from a mail order company, it is still worth doing your next stage of research in a shop where you can actually see and handle different models. This is particularly important for getting the feel for a particular model. As described in Chapter 1, buying a camcorder is a bit like buying a glove. It must feel comfortable in use, the main controls should fall easily to hand, and its size should suit your needs. Therefore you should make a visit to a shop that specialises in selling camcorders – which will have a good range of models on display and which are wired up to televisions so that you can see what you are shooting.

Get into a routine when comparing different models of checking things such as how quickly the camera focuses when moving from subject to subject. You should also put a tape in each camcorder, record some footage, and replay that to check the true recording quality of the camera. The 'live' picture a camcorder feeds direct to a television can look much better than the recorded footage. Any good camcorder shop should allow you to do this – though you may have to provide your own tape (a small investment to make).

Many a harsh word is spoken about shop staff who either know nothing about what they are selling you or who bamboozle you with jargon. You cannot, however, judge the quality of service from the outside of a shop. A small town shop is not necessarily going to be better at helping you than a large store which is part of a national chain. It is all down to luck as to which salesperson you get to speak to. Even if they have some knowledge about camcorders, they are probably not experts. You should therefore form your own opinions, and not rely on getting good information from the shop.

Added incentives

In order for stores to make their camcorders look more attractive, it is common practice to package them with 'free' gifts. A video light, a tripod, a handful of tapes . . . all these and more may be offered as incentives. Indeed, enticing offers such as '£200 worth of free accessories' may seem hard to refuse. It is worth pointing out, though, the reason why stores produce these package deals. The first, and most important, is that the store is likely to have a much higher profit margin on accessories. It is therefore more economical for the store to offer a £100 free gift, than knock £100 off the price. Secondly, it makes it much harder for you to tell whether you are getting a bad deal or not – as you cannot make direct comparisons with the price in other shops.

You should therefore estimate how much these promotional items are worth to you – if you have an old tripod, or do not like the free bag, then do not count the value of these in your comparisons.

You should also check that the camcorder includes all the accessories that you will need immediately in the price. You should expect:

- a rechargeable battery
- a battery charger
- a cable to connect the camera to the mains via the charger
- a VHS-C cassette adaptor (if buying a VHS-C or S-VHS-C camera)
- an AV lead, for connecting the camera direct to a television or VCR
- an RF lead and converter, for connecting your camcorder to the aerial socket of a television (not all TVs have AV sockets, so an RF unit is often essential). As this is quite an expensive item it can often be left out to cut costs.
- a tape
- a shoulder strap
- a lithium battery (for the camera's internal clock)
- a lens cap
- a shoulder bag (fast becoming the exception rather than the rule)
- an instruction book.

If the has one of these items missing, you may be able to persuade the retailer to throw them in for the price anyway. An extra battery or tape might also be worth negotiating.

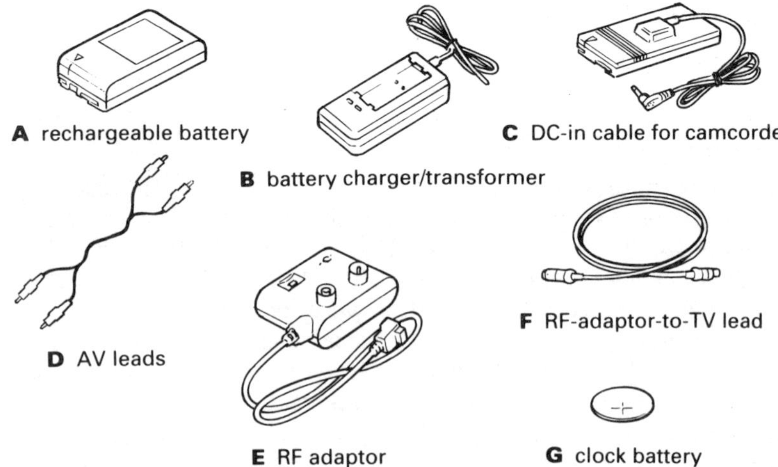

The essential accessories that should come with your camcorder

Grey imports

One of the most contentious issues in photographic and video retailing is grey importing. A grey import is one that has been brought into the country by someone other than the official distributor for that country. The phenomenon exists because the same camcorder can be cheaper in some countries than others. It is therefore worth the dealer's while, say, to buy his stock from a wholesaler abroad, rather than going to the official source for his country. He can then pass some of this cost saving to the consumer.

As any particular model of camcorder is likely to have been made on the same Japanese production line wherever it ends up being sold, the consumer, in theory, should have little to fear. However, the manufacturers generally do not take kindly to this kind of international trading – and they have one powerful weapon to help them discourage it. Commonly, the warranty is only valid in the country where the camcorder was originally supplied. So if you buy a grey import, you are unlikely to get an official manufacturer's warranty. To make up for this, the dealer will normally offer his own warranty on the same terms as the manufacturer, but some people find this disconcerting.

Buying abroad

As the price of camcorders varies around the world it might seem an attractive idea to do your shopping while on a trip abroad. However, doing so could prove a costly mistake – as it could be completely useless when you get back home. The main reason for this is that there are three different television formats commonly used around the world – and none of them is directly compatible with the other. So if you bought a camcorder abroad it would work fine – until you tried to play back your footage on the television, or tried to copy it on your home VCR – you could end up with no picture or sound at all.

In the UK, the television standard used is known as PAL (Phase Alternate Line) and this is shared with a large number of European countries, Australasia and large chunks of Asia. The United States and Japan use the NTSC (National Television Standards Committee) system. Finally, SECAM (Séquential Couleur à Mémoire) is used in France and parts of the old Soviet Bloc.

TV standards around the world

NTSC	PAL	SECAM
USA	UK	France
Japan	Ireland	Saudi Arabia
Canada	Germany	Zaire
Greenland	Scandinavia	Lebanon
Taiwan	Singapore	Iran
Barbados	Zambia	Iraq
South Korea	Tibet	Albania
Mexico	Netherlands	Poland
Philippines	Spain	
	Australasia	

To make matters worse there are small variations in the standards themselves. Add to that the differences in mains electricity supply, different warranties, and import duty, and you should hopefully be put off the idea of shopping overseas. Unless you know exactly what you are doing, the risk is not worth the money you could save.

Buying secondhand

Buying a camcorder secondhand might also sound like a good way to get into the world of video making without spending a fortune. But although cars are a good buy once used, the same does not always hold true for camcorders. The real problem has been the falling price of new camcorders over the last few years. A camcorder that cost £1,000 three years ago might have fewer features than one that costs £600 today. But it is unlikely that the person selling the three-year-old camcorder is going to take this fact into account when deciding on a price. Even if the price is low, you must be absolutely sure that you are getting all the accessories you need (see above) – and you must be sure that the machine is in full working order. If you buy secondhand from a dealer, you are not going to get the same length of guarantee as you would when buying new.

As a rough rule when buying secondhand, you should expect that a camcorder is worth only half its original price as soon as it has been taken out of its box. You can then take another 20% off this for each year that it has been used. If you can get this sort of price, have made sure that you have all the requisite accessories, that it looks practically brand new, and know that the camcorder is unlikely to break down as soon as you use it, then you might have a reasonable buy. If not, stay away. Checking for faults on a used camcorder is not a job for the untrained eye. Unless you are very lucky, it is far better to hunt around in the shops for special clearance deals than chance your arm in the secondhand market (see the section on last year's models below). On the other hand, if you are selling an old camcorder, sell privately, make sure you hand over all the accessories, and give the best price that you can afford.

Last year's model

The average camcorder model is likely only to be sold for twelve months before it is replaced by its manufacturer. Top-end camcorders might last twice as long, but fast-selling budget models may only be on the books for just six months before being upgraded. This state of affairs provides the best opportunities for the bargain hunter. The camcorder is sold as new – so you get the full outfit of accessories and full manufacturer's guarantee – but you pay, perhaps, only 50% of the original launch price.

Buying last year's model does not necessarily mean that you are getting a camcorder that is out-of-date. Some so-called upgrades do little more than a little cosmetic surgery to the original model and add a few gimmicky features that you might well do without. What you will certainly be able to get is more creative features for your hard-earned cash. While on the current range you might only be able to afford the bear minimum of features, by looking at older discounted stock you may find you can stretch your budget to high-band quality, stereo sound and a whole host of other features. Some words of warning though, don't rush into your purchase because the dealer seems to be offering an incredible bargain. You must still shop around and do your research – there might still be a better buy around the corner.

Warranties

There are two main reasons why a camcorder is likely to need repair. The first is that there is something wrong with the camcorder when you buy it. The second comes from misuse – such as the owner prodding the insides of the camcorder with a screwdriver, or getting sand and water inside the casing. Although a guarantee will not help you with the latter, it will help you if you are unfortunate enough to suffer the first. The most important thing you can do to get the best value out of your warranty is to use your new camcorder as extensively as possible as soon as you buy it. In this way, you can check out all its facilities, and see if they work correctly. If something is wrong, you will be best advised to take the problem to the shop where you bought the camera in the first instance. They will be able to check whether a repair is necessary, and can advise you on the procedure for getting the warranty honoured.

Extended guarantees

Many dealers will offer you the chance to buy an extended warranty lasting anything from an extra one to four years. True, if something does go wrong with a camcorder, it can be an expensive business. Camcorder repairers need to be extremely skilled to work with the miniature circuitry in a modern camcorder – and this means labour is expensive. Parts can be expensive too, as more and more of the circuitry is put on fewer and fewer circuit boards – which may well need completely replacing when a fault develops. However, with any extended guarantee, you are taking a gamble on the odds of something going wrong after the usual warranty time. In some stores the profit made on selling an extended guarantee can be as high as on the camcorder itself – suggesting that the odds on this particular wager are not stacked in your favour.

Mail order

Some people would be horrified by the idea of buying something as expensive as a camcorder through the post. Buying mail order, however, can have its advantages – especially if you know exactly what you want. The most obvious benefit can be price. Not only can you extend your search to bargains much wider, but mail order outlets often have lower overheads and can pass on cost savings to you. What you lose is the personal contact with the salesman, and the ability to check the goods before you buy. If you do buy mail order, there are ten golden rules that you should follow:

1. Always check exactly what the price includes. Does the price include tax and carriage? What sort of warranty are you getting, and what accessories are included?

2. Keep a full record of your telephone conversations, including the date and the name of the person you spoke to.

3. When placing an order, confirm the details in writing by post. Use recorded delivery post, or similar, so that you will have proof of receipt of your letter should things go wrong.

4. Keep copies of all correspondence.

5. Check the company's policy on refunds or cancellations. Once you have placed the order you have made a contract to buy. Changing your mind is not reason enough to get a full refund.

6. Always stick to your guns. Mail order firms may advertise an incredible price on an item, but you find they have sold out. They then try to sell you another item. If this is not what you want, try somewhere else.

7. Check if the item they are offering you is in stock. Mail order companies may take your order, and then wait for delivery from the manufacturer. This can delay your order for weeks. State in your conversations and correspondence that time is of essence to the contract – so you have made it a condition of purchase that the item is despatched promptly.

8. Pay by credit card if possible. This should speed up delivery (cheques will be cleared before your goods are despatched). Credit card payment may also give you free insurance or other worthwhile benefits – you may also get valuable help from the credit company should things go wrong. Insist, however, that your card is only debited when your goods are ready for despatch – otherwise you could pay before the company even takes delivery of your order.

9. If things go wrong, contact the publication where you saw the advertisement. They may be able to offer you advice and help on what to do.

10. As with any purchase, check your goods thoroughly as soon as you receive them. Use them as fully as possible, as quickly as possible – so if something is wrong you can let the mail order company know as soon as is reasonably feasible.

6
— THE BASICS FOR — SHOOTING

This chapter is for the impatient reader. You have your camcorder – and you want to start shooting straight away. Look out Steven Spielberg, here you come! Of course, you should begin by reading the instruction book before you start guessing how to get your camcorder up and running. Although there are lots of similarities between different models of camcorder, there are important differences in how things are done – and this is where any instruction manual comes into its own. However, most people have a pathological fear of these things – acquired from years of frequently finding them overlong, overshort, or just downright incomprehensible. This chapter is not meant to replace the instruction book, but is designed to get you shooting in the minimum amount of time and with the minimum amount of damage.

Getting powered up

However much enthusiasm you have for getting your first shots on video, you will need power. The first thing you should do is charge the battery pack – as this will be in a discharged state when you buy it. This will take you about an hour – giving you enough time to digest the information in this chapter.

First the charger unit must be fitted with a mains plug – if this has not been fitted in the factory. If using a fused plug fit a three amp fuse. Then slot the battery on the charger – so that the metal contacts on each meet. Switch the power on at the mains. A light will appear on the charger and your battery will now be charging. Most chargers have a light telling you when the process is complete.

For those who do not want to wait an hour, you can run your camcorder direct from the mains straight away. You will need the charger, plus an additional wire that connects between the camcorder and the charger. This wire will plug into a socket marked 'DC out' on the charger, and then will connect to the camcorder via either (a) a socket marked 'DC in', or (b) to the battery terminals of the camcorder using a dummy battery connector. The disadvantage of this quicker route is that most chargers will not allow you to charge the battery at the same time as you are using them as a transformer to power the camera directly.

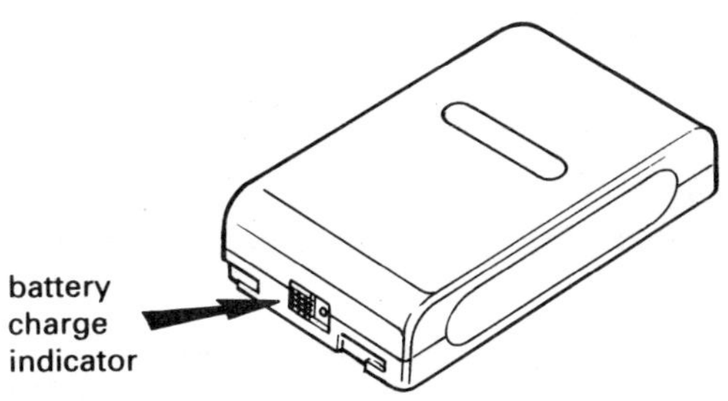

A battery charge indicator has to be set manually

So that you can tell whether a battery is charged or not, most nicads have a marker switch. You should move this so that the red mark is showing when you have charged it – and move it to its other position when the battery runs down. As long as you remember to use the switch, you will always be able to tell if a particular battery is powered up or not.

While on the subject of batteries, there is a separate power supply to fit – and that's the small button-shaped cell used to power the camcorder's internal clock. The power supply is separate, so that you don't have to keep resetting the time and date on each occasion you recharge the main nicad battery.

Inserting the tape

To put the tape in, the camcorder must have power, and the main control has to be switched on. By pressing the eject button, the tape door will slowly open automatically – do not force it to go any quicker. The tape will only fit in one way. The easy way to remember which way round the tape goes is that the hinged flap at the top of the tape goes into the machine first, with the transparent tape window facing the outside of the machine. To close the compartment you gently press the door where indicated (usually by the word push). There will be a whirring noise as part of the tape is loaded on the transport mechanism. You are now ready to shoot.

Recording

One of the most fundamental lessons you must learn about the camcorder is how to get it into record. This might seem obvious, but it is remarkably easy to record when you think the machine is switched off – giving you the dreaded footage of the ground as you walk along with your camcorder at your side. Even more tragic is to think you are recording when in fact the camcorder is off – so unrepeatable moments are lost.

The camcorder, in fact, has three main modes: off, pause and record. To turn the camera on in the first instance, you will find there is a main control switch, normally marked 'operate' – this often has an accompanying light to show you when it is in the on position. If you want to record, there is a second switch you may well have to adjust too – the one that changes the camcorder from camera to VCR, depending whether you want to record, or want to play back a tape.

Once the camcorder is switched on, and in the camera position, the viewfinder will come on – showing you exactly what the lens is seeing. This is the pause mode – and you will see the word 'pause' displayed somewhere in the viewfinder. This mode is designed so that you can frame your shots up before you start recording. To start recording from here, it is just a simple matter of gently pressing the red start button – which usually falls readily to you right thumb. Press this and you start recording – press it again and you return to pause. The only differences between these two modes are either having the words 'record' or 'pause' in the viewfinder, and a slight increase in noise from the camcorder, as the tape starts to move in record mode (on some camcorders pause is called standby, or 'stby' for short). Practise swapping between both modes, so that you are completely at ease with which one you are using at any given time – this could help avoid costly mistakes in the long run.

If you cannot read the words clearly, you may not have the viewfinder

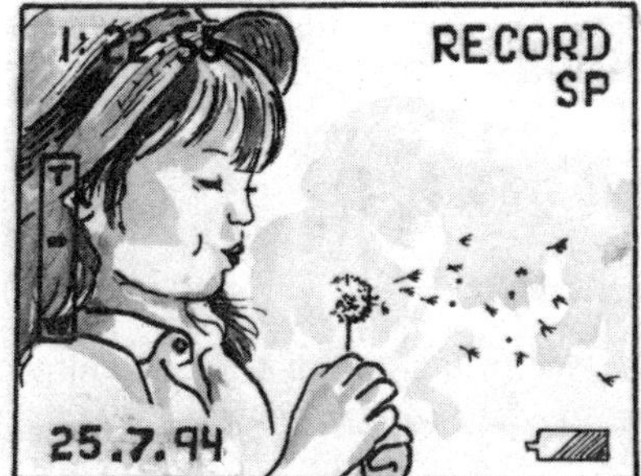

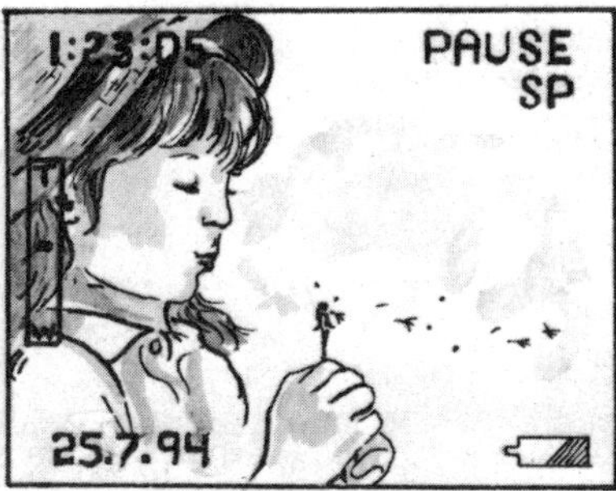

A typical viewfinder display – packed with information. The most important things to pay attention to are the words RECORD and PAUSE

adjusted properly. The end of the viewfinder is basically a magnifying glass that is focused on a television screen below – the distance between the TV and the magnifier will depend on your eyesight. First ensure that the viewfinder is fully extended – as some can be pushed in to make them shorter when not in use. Then twist the small ring behind the rubber eyecup until you can see the words in the viewfinder clearly. Do not use the picture on the screen to make this adjustment.

Backspacing

It is important to realise that when you switch from pause to record mode the camcorder doesn't start recording instantaneously. In fact, it can take a couple of seconds for the camera to wrap the tape round the recording heads and get up to speed and correctly synchronised. To allow for this, when you switch back to pause, the camera will wind back the tape slightly – this is known as backspacing. You need to know this, because it means that you must always allow a second or two between pressing the record button and the recording to start – and you must also allow a second or two of extra footage to be recorded at the end of a particular sequence, before you switch to pause. This process can mean that it is hard to switch on quickly for fast-moving events – as you need a few seconds to prepare for them. If you want to end up with a short sequence of a golfer's swing – you must start the recording as he addresses the ball to ensure that you are recording before the golfer starts his backswing. Then when he reaches the top of his follow-through you must keep on recording for a couple more seconds to ensure that you do not miss the end of his swing. The exact amounts of time that you need to allow depend on your camcorder – and are well worth familiarising yourself with if you do not want accidentally to cut off valuable footage.

Holding your camcorder steady

A shaky picture is one of the most common faults you will encounter when making videos. It is therefore important to hold your camcorder as steadily as possible at all times. The main support for most camcorders is the wrist strap that fits around the right hand. This should be tightened using the Velcro fastener around the four fingers – leaving the thumb free to control the record button. On bigger camcorders extra support is given via a padded recess at the bottom of the back of the camera that rests on the shoulder. Although it is possible to hold the camera with one hand, your left hand should also be used whenever possible to give extra support. This should either clasp the left side of the camcorder, or should cradle the underside of the lens (if this protrudes from the main camera body). The lighter your camcorder, the tighter you'll need to grip. You'll also notice the effects of camera shake more the more you zoom towards the telephoto end of the lens.

Camera steadiness can be further improved by the way that you stand. In normal situations you should stand up straight with your feet slightly apart – so that they are directly below your shoulders. You should tuck your elbows into your body.

Obviously, for the steadiest of shots you need to use a tripod. Unfortunately these are cumbersome pieces of equipment that few are willing to cart around with them. Instead, look for other ways of giving your camera extra support. Here are just a few examples:

- Stand against a door frame, a tree or lamp post – pushing your shoulder into to it for extra support.
- Sit on a chair backwards – propping your elbows on the back of the chair.
- Lie on the ground – this time propping your elbows on the floor to support the camera.
- Sit with your back against a wall – resting your forearms on your knees.
- Go down on one knee – resting your elbows on the raised knee.

Wherever you are, you will normally find something, or some position, that will give you extra support, without ruining your angle of view.

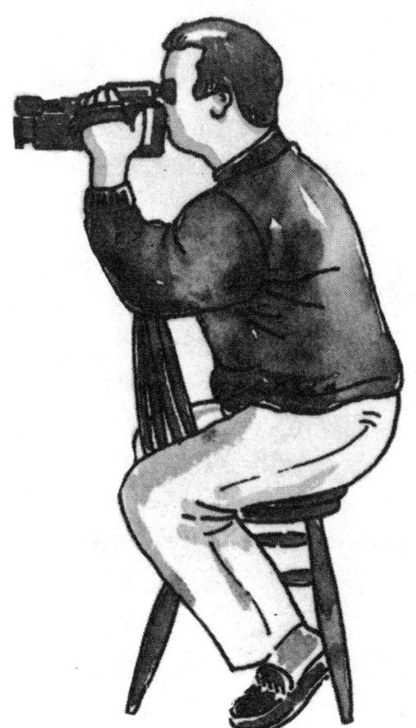

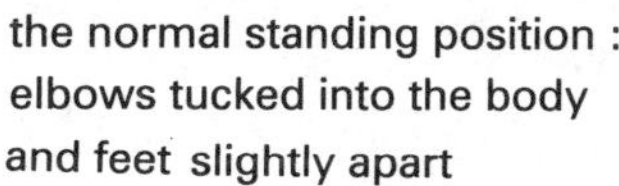

the normal standing position : elbows tucked into the body and feet slightly apart

using chair for added support

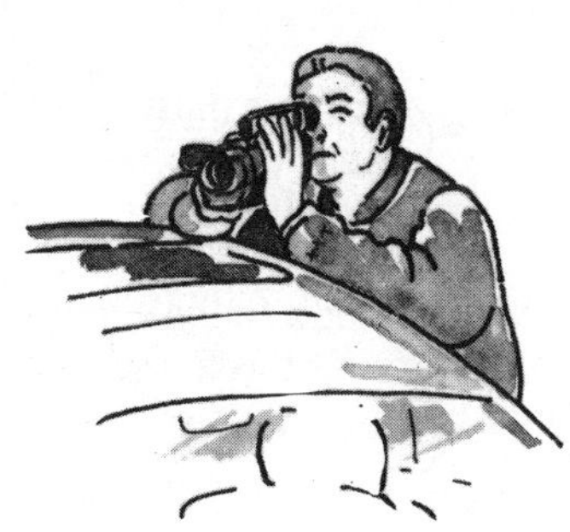

using car for added support

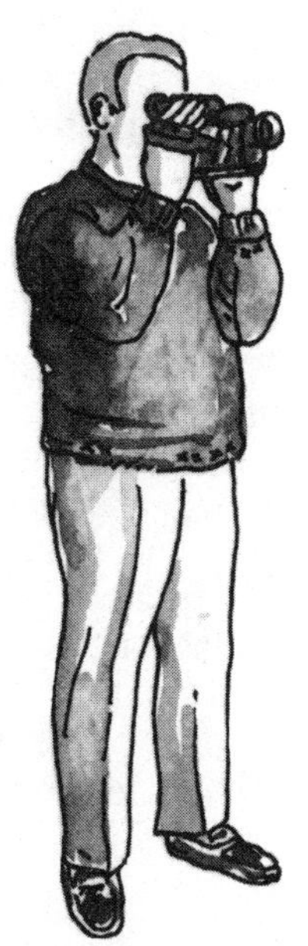

lying on ground for worm's eye view

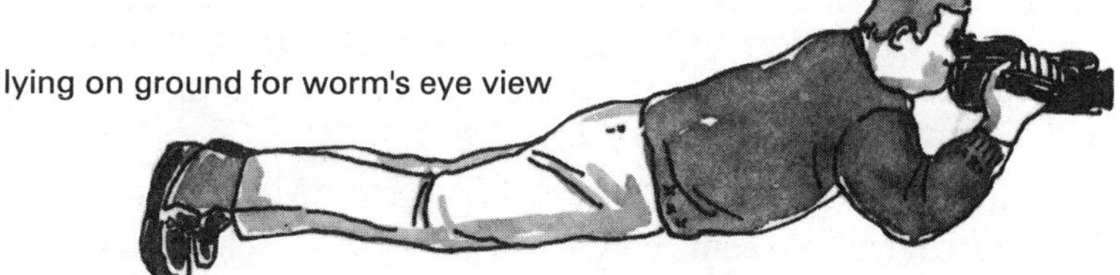

Two minutes on technique

Most of the rest of this book is devoted to teaching you about how to use your camcorder. However, if you want to get shooting right away, you could do far worse than trying to remember these five simple pointers to successful video making:

1. The motorised zoom is meant for framing up your shots before shooting. You should only zoom while you are recording in exceptional circumstances. You should never zoom in and then out again, or vice versa, while recording.
2. You should not keep the camcorder recording without pause for minutes on end. A good video is made up of lots of brief shots – perhaps only lasting 10–20 seconds each. Remember less is more – no one wants to watch a 30-minute video when a 10-minute one would have done just as well.
3. Do not swing the camcorder around from subject to subject while recording. Record one subject, pause, move the camera round and frame the next shot, then press record . . . and so on.
4. Add variety to your video by moving around between different shots – take some shots from close-up, and some from a distance, take some with the wideangle end of the zoom, and others with the telephoto. These visual changes will make your videos look more interesting.
5. Remember that you are telling a story – so you must keep the viewer informed of what is going on. If your subject laughs at something, show a shot of what they are laughing at. If you are about to film at a new location where you are going to use your camcorder (an historic castle for a day out, for example, or a friend's house for a barbecue), stop to take a brief introductory shot from the distance before you get inside – so the viewer has a point of reference for what is to follow.

Connecting up to your TV

Although most camcorders allow you to watch what you have shot through the viewfinder, the quality of the playback through such a small monitor leaves a lot to be desired. A television is therefore the camcorder owner's most important accessory.

The thought of having to wire your camcorder up to your television can seem frightening to some people. What lead do you plug into where?

What channel do you turn your TV to? The confusion is exacerbated because the sockets found on camcorders, and televisions, vary from model to model. It is hardly surprising that some people choose the VHS or VHS-C formats so that they can replay their tapes through their VCR – rather than have these wiring worries.

The best way to link your camcorder to a television is by using direct AV (audiovisual) connections. However, older and smaller television sets do not have the necessary AV input sockets for this. With these sets the only way you can get the picture and sound into the TV is via the aerial socket. To this end, most camcorders will come with an RF converter – a device that is necessary to convert the output signal of your camcorder into a form that can be input into the aerial socket of your television.

An RF (radio frequency) converter is basically a box that transforms the electrical signal from your camcorder to the same form as a television channel arrives from the television's aerial. This signal is then reconverted back by the television. This conversion process involves a loss of picture quality, and therefore should only be used if you cannot use the direct connection root.

The RF converter comes attached to a wire which terminates with a set of plugs or pins that will fit the AV out socket (or sockets) on your camcorder. On the RF converter there is a standard aerial plug. This will not fit directly into the aerial socket of your television, but needs a separate aerial extension lead to do this (again usually supplied with the camcorder). The aerial lead you have removed from the back of your television (the one connected to your aerial!) can be connected to a socket on the RF converter – this is optional, as it is only necessary if you want to be able to switch quickly from TV to camcorder, or vice versa.

Once the connections are made, the next step is to tune the television into the signal from the RF converter. The simplest way to do this is to find a channel on your TV that is not used (preferably one that does not receive any TV station when the aerial is connected). The RF converter is then tuned into this blank channel. This is done by turning a small screw (using a small screwdriver) on the RF converter – this is typically marked something like 'Ch. 30–40'. Turn your camcorder on so that it is in pause mode with an image in the viewfinder. You then adjust the screw until this same image appears on the television. And, finally, you are ready to watch your home movies!

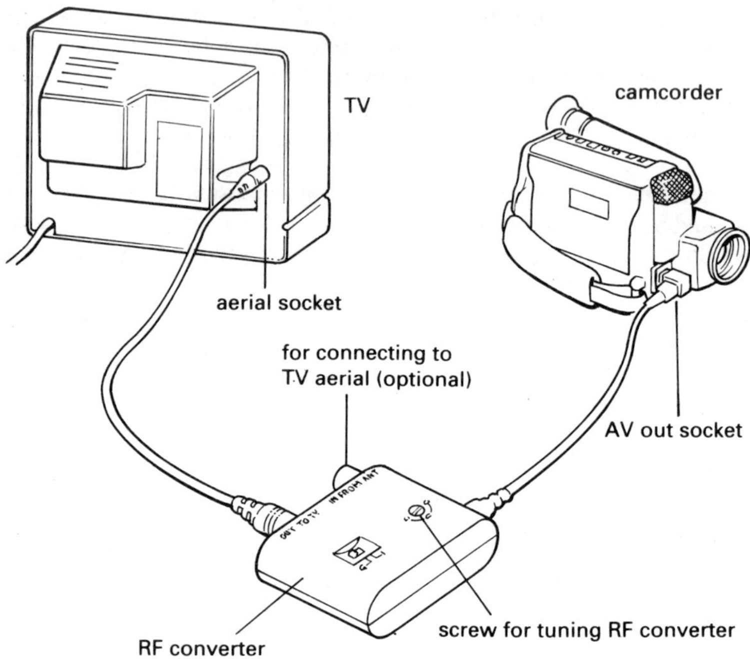

How to connect a camcorder to a TV using its aerial socket

Fortunately, making direct connection to your television is much easier – once you have the right lead. Most camcorders are supplied with an AV lead with plugs to fit the AV output of your camcorder at one end, and two or three phono plugs at the other (one for video, plus one or two for sound, depending on whether you have a stereo camcorder or not). This is fine if your TV has the appropriate phono socket AV inputs. Unfortunately this is no always the case. Some TVs will only have Scart inputs (or even some other form of socket) – in which case an extra lead may have to be purchased.

Once you have wired up directly in this fashion, to get the picture on screen you will have to switch the TV into AV mode (using the appropriate button on your remote control or TV itself – usually marked AV). You may have to press this button more than once if your television has several input sockets, so that you can find the right one. Switching your camcorder into record pause mode, as before, will help you achieve this.

Finally, if you have a high-band camcorder (using the Hi8, S-VHS or S-VHS-C formats), you should use slightly different connections. You can use either of the above two methods, but neither will allow you to see the full advantage of the high picture quality that is possible with your equipment. For this you must use the S-Video sockets on both your camcorder and TV. Not all TVs have this socket, but if your TV does have one use it – the necessary lead will be included with your camcorder. As this lead only sends the video signal, the audio signal must be connected separately (normally using one or two phono plugs).

Although the connections between the TV and camcorder can be complicated at first, once you have identified the correct wires for your set-up, and had them working once, you will find that it is a simple procedure to repeat in the future.

7
— LENS CONTROLS —

It is sometimes said that a video allows you to see exactly how the world is without having to leave your armchair. Although the camcorder can provide a lot of information about what the things it is pointed at look like, it does not tell the whole story. One of the main reasons for this is that the camcorder's lens is not like the human eye. We cannot see something and zoom in on a close detail, masking out all around. A camcorder, on the other hand, can do more than this. If there is

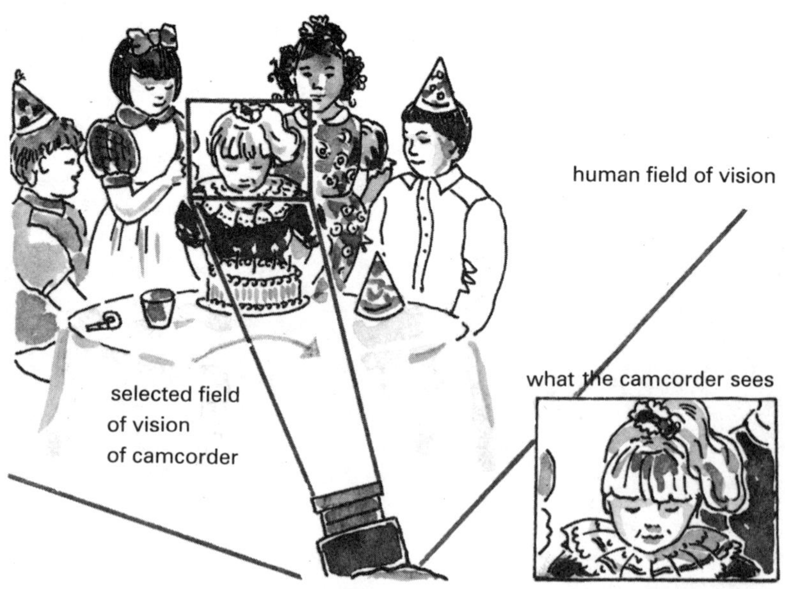

The camcorder is selective in what it sees

something in the foreground or background that the camcorder operator does not want the viewer to see, it can be thrown out of focus so that it becomes an unrecognisable mass.

In short, the camcorder is selective, not only in what it shoots, but how it shoots it. The human eye sees all before it – not just what we want to look at, but everything that surrounds it. A camcorder, in a trained pair of hands, will see only what its owner wants it to see.

Zooms

Nearly every camcorder that is made has some form of zoom lens built into it. The key advantage of this remarkable set of lenses is that it allows you to get 'closer' to a subject without having to move your shooting position. In fact, each time you double the focal length of the lens, you double the size of your subject on video. With a 10x zoom, a subject that is shot with the wide setting so that it formed an image just one inch high on the TV screen, can become a ten inch high image just by using the longest telephoto setting of the lens – all without moving your feet.

Although using the more telephoto settings on your zoom lens can give impressive results, it is not a good idea to use them without good reason. For the most natural looking videos, you should try to use the wideangle lens of your lens whenever possible. The main reason for this is that this setting most closely approximates to the normal field of vision of the human eye. To check this for yourself, look through your camcorder at around its widest setting keeping your other eye open – you'll notice that the subject in the viewfinder is about the same size as seen directly. Because of this, scenes recorded with this focal length (or one close to it) will have a normal-looking perspective (see below).

Another reason for keeping the focal length short – and therefore the shooting distances to a minimum – is that although light travels great distances without any distortion, the same cannot be said for sound. With your powerful telephoto lens you may be able to home in on a single face on the other side of the street – but you will not be able to record what they are saying. Sound, especially that of human voices, does not travel well and soon becomes muffled and unrecognisable. We will be looking at the ways of getting round the problems of recording sound in chapter 12.

Perspective

Although zooming the lens is frequently described as getting 'closer' to the subject it does not give the same visual effect as moving forward with the camera. The reason for this is that the perspective of a shot remains the same if you stay in the same position – whatever you do with the lens. If you move forward, however, the perspective will change. Perspective is directly related to subject distance and nothing else.

In practice, however, perspective does have a lot to do with the focal length of lens that you are using. This is because telephoto lenses are generally used at greater distances away from the subject than wideangle ones – for obvious reasons!

A telephoto lens, when used from a distance, tends to compress perspective, making subjects that are in fact hundreds of yards apart look as if they are practically sitting on top of one another. If you shot the same scene with a 'normal' lens (the wideangle setting on most camcorders) from the same distance you would find that the perspective was exactly the same. However, because the telephoto lens only sees part of the 'normal' view, the perspective looks unnatural. The greater the telephoto setting you use, the greater this apparent compression becomes. The reason for the effect is that in reality all the subjects in the viewfinder are some distance away from the camcorder – but because you are only seeing part of the scene, they are presented as being very close. If they really were close they would differ in scale – so similarly sized subjects that were closer to the camera would appear bigger than ones farther away.

There are practical advantages to this effect, however. For example, you could, from the right vantage point, shoot the Sacre Cœur and Eiffel Tower in Paris in the same frame, without one of them being a small dot in the distance.

A wideangle lens, on the other hand, has an opposite effect on perspective. Unfortunately, few camcorders have a wideangle setting that gives a significantly wider view than that which is considered 'normal' by the human eye. However, wideangle converters can be bought for most camcorders that stretch the angle of view of your camcorder's widest lens setting. A 0.5x converter, for instance, will double the maximum angle of view (halving the minimum image size possible from any particular distance). With a true wideangle view, therefore, you can add depth to a particular shot. Objects in the distance appear

wideangle

telephoto

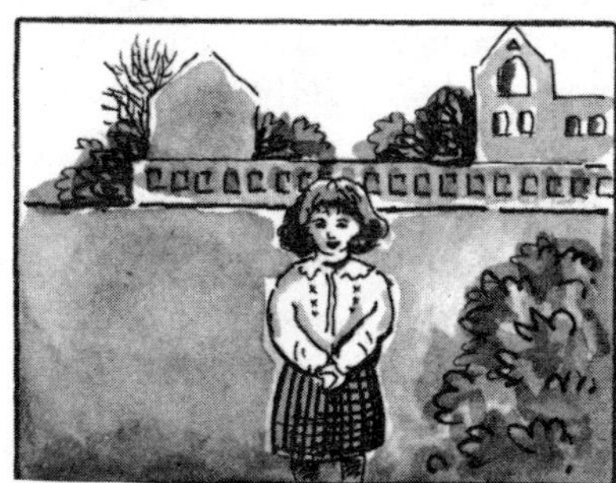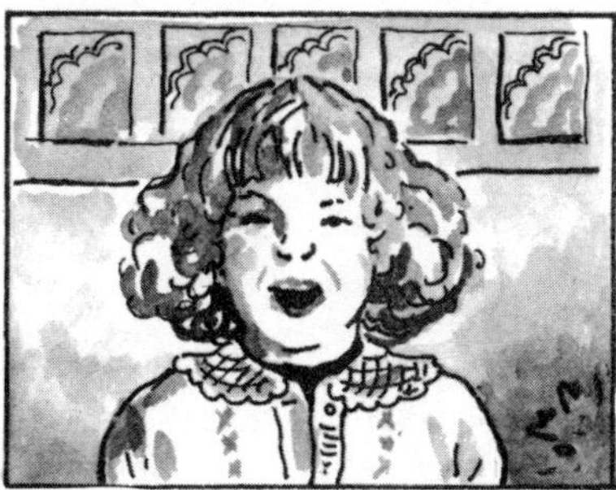

zooming in closer

The main subject is enlarged in the frame – but so is the background, making it seem much closer to the subject

wideangle

wideangle

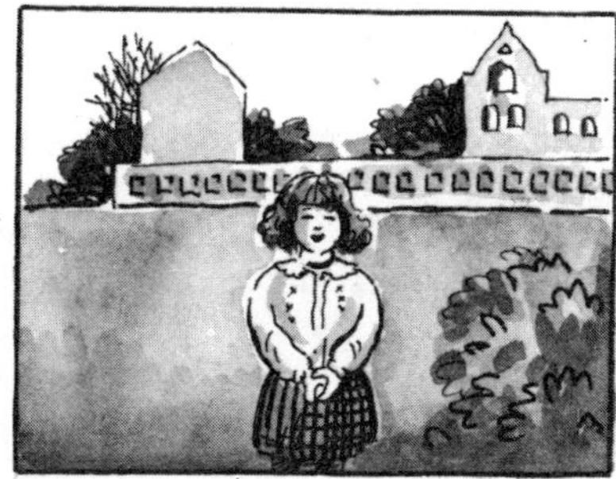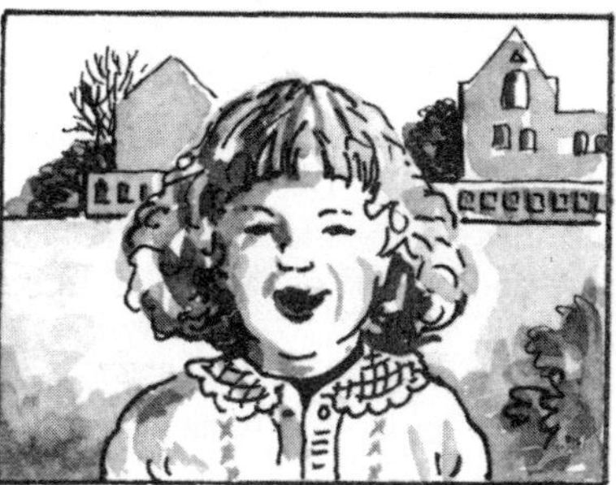

moving camera in closer

The main subject is enlarged, but the background remains very similar to before, so now looks less dominant in the picture

smaller – and therefore seem further away – than with a normal lens. Subjects in the foreground, on the other hand, seem to dominate the frame. Everything in the picture seems to be much more spaced out than it is in reality.

In the real world, the focal length that you choose for a particular shot will often be forced upon you by circumstances. You have to use your most powerful telephoto setting to shoot wildlife – because if you go any closer you run the risk of it moving away, and missing the shot altogether. If shooting a great cathedral, it may be impossible or impractical to move farther away from it to be able to use a longer focal length. However, there will be plenty of other occasions, especially when videoing people, when you will be able to have a free choice over focal length and subject distance. In these circumstances use a normal view for most of your shots – only using more telephoto settings when they are really justified.

Camera shake

Another good reason why you should favour the wideangle setting on your zoom lens is that it becomes more and more difficult to hold the picture steady the more you increase the image magnification. Imagine that the involuntary movements of your body while you are holding the camera mean that the camera is moving by half a degree from side to side (or up and down). If you are using the wideangle setting of your lens, which has, say, a 50° angle of view – this small movement has little effect on the picture. If you are using the telephoto lens on a 10x zoom, however, which has, say, a 5° angle of view, this movement represents a very shaky picture.

Of course, these problems can be practically eliminated by using a steady tripod. But today's highly portable camcorders mean that we are less inclined to take such a cumbersome hunk of metal around with us – and built-in image stabilisation systems are becoming commonplace on camcorders.

Electronic zooms

There is no doubt about it, a 100x zoom sounds appealing. Imagine being to be able to get a decent sized image of a robin from 50 yards away. Unfortunately, zooms that offer such image magnifications produce the effect by electronic magnification. In reality, a so-called 100x zoom produces an image with a normal 10x zoom and then blows up the central portion of this image to fill the complete frame. The result is not just a poor quality picture, because instead of using 370,000 picture elements to give an image, only 3,700 are being used. When these are enlarged to fill the whole frame you find the picture is made up of visible squares rather than practically invisible points – each square being made up of 100 pixels of exactly the same colour and tone. Nearly all the detail in the picture is therefore lost. These zooms, therefore, only have any use when shooting a simple, bold shape, that is still recognisable after all this manipulation. Even then the image doesn't bear too much close scrutiny. It will give you a good full-frame shot of a full moon – but do not expect to be able to make out any craters. Again the problem is made worse by camera shake. Such a small angle of vision means that even a tripod will have trouble keeping the shot steady if there is a breeze blowing, or a car drives past. In fact, these monster electronic zooms could serve well as early-warning systems for earthquakes!

—— The true purpose of the zoom ——

Because any camcorder zoom gives such a powerful telephoto setting, it is easy for camcorder owners to get carried away with its capabilities. They zoom in on something in one part of the frame, zoom out, and hunt around for something else to zoom in on again. In and out, in and out, like a trombone it goes – great for personal amusement, but if this is recorded it will make anyone who has to watch it feel ill.

One of the best ways to learn how to make good movies is to watch television and see how the professionals move from subject to subject with their cameras. One thing you will soon notice if you do this is how rarely you see the zoom lens being used. Instead you will see a continuous series of shots, all taken with different focal lengths – some wideangle, others telephoto, and the rest somewhere in between. So if the zoom lens is not used during recording, what is it there for? The true purpose of a zoom is to frame up your shots before you press the record button. Trombone the zoom back and forth while you are in record pause to look for the perfect shot, but when you've found it, take your finger off the zoom control button and record it for a few seconds. If you then want to zoom in to take a closer view, do so – but in record pause – and then record again. The results you get will be far more easy for the viewer to watch.

Because of the construction of many of today's zoom lenses, nearly every camcorder has a motorised rocker switch to control the zoom. The manual zoom ring or lever, with which you physically extend or shorten the lens to the required focal length, is becoming a less common feature. If your camcorder has one, it is well worth using instead of the motorised control. For one thing, it can be quicker to move it from one end of its range to the other. Most importantly, however, using the manual control means that you will not be using up valuable battery power. Save the motorised control for the rare occasions when it is legitimate to zoom while you are recording – as it will give a smoother-looking effect than zooming manually.

There are three occasions when you may need to zoom while recording:

- *Subtle compositional changes.* People are unpredictable subjects to video. You can never be sure exactly what is going to happen next. You might be videoing one member of the family, and then another sits down beside them. To fit both of them in you may need to zoom out slightly to fit them both in. Such small changes in the focal length during shooting are acceptable – but you should make them as slow as possible so that they are not noticeable to the viewer.
- *Creep zoom.* Sometimes it can be useful to zoom in on a subject to reveal a detail that is not noticeable in the longshot, for dramatic purposes. You might start a murder mystery movie with a shot of an ordinary-looking street – and gradually zoom into a shot of a shifty-looking character standing on the corner. The secret of doing this successfully is to make the zoom last as long as possible – so it creeps along its range.
- *Crash zoom.* This is the exact opposite of the creep zoom, and again used for dramatic effect. Here the zoom is moved as quickly as possible from a wide shot to a telephoto close-up – normally to reveal the facial reaction of a character to something that has just happened. Use this technique very sparingly.

Depth of field

One of the most important controls that you have at your disposal when using a camcorder is to decide exactly which parts of the picture are in sharp focus. Unimportant or distracting parts of the scene can be thrown so far out of focus that they become unrecognisable – concentrating the viewer's attention on what you want them to see, and nothing else.

In strict scientific terms, a lens can only focus sharply on one plane at a time. You focus the lens on one particular point in the frame (or the camera's autofocus system does it for you) and only the parts of the picture that are the same distance away from the lens are critically focused. Anything nearer or farther away from the lens becomes increasing out of focus the more distant it is from this plane.

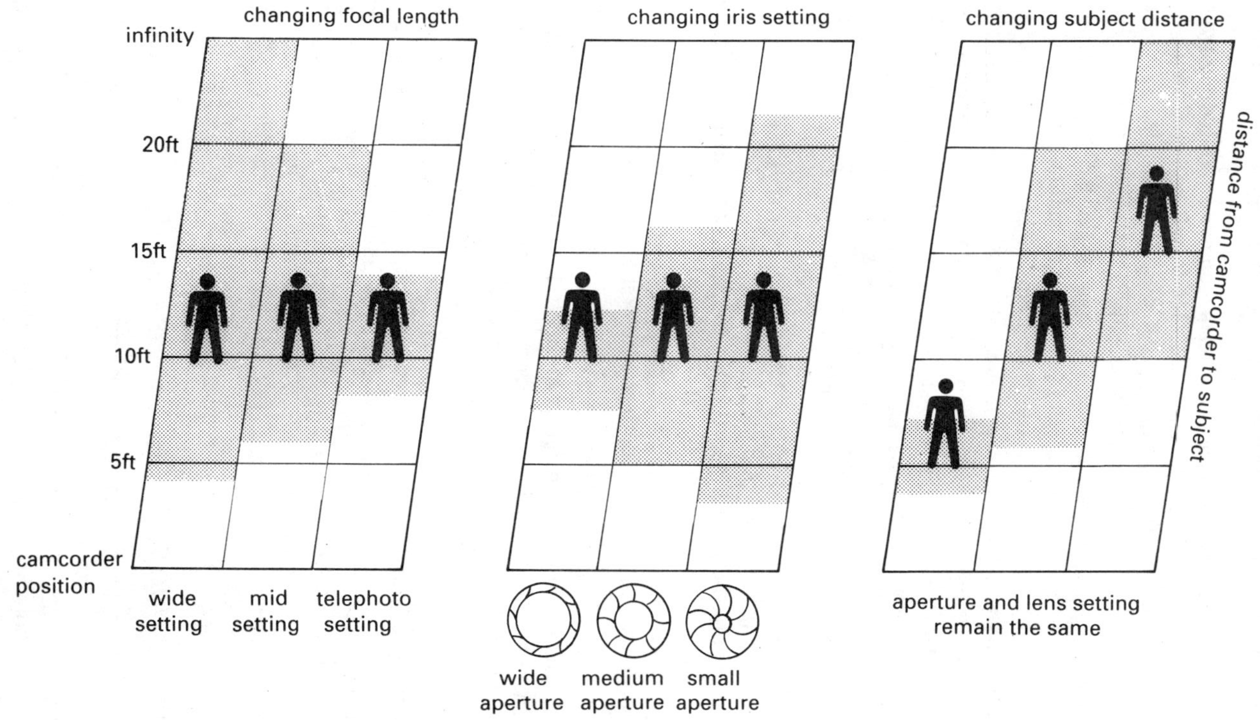

How the depth of field is affected by altering zoom setting, iris and shooting distance

In practice, however, there is a reasonable amount of latitude in what we actually see as sharp in the picture. There will be some things in front of the critically focused plane, and some behind, that will look acceptably sharp. What is in fact happening is that points on the plane of critical focus are focused as points of light on the CCD chip. Points that are not on this plane are resolved as larger and larger circles on chip, depending how far they are away from the plane. But our eyes will accept a circle up to a certain size as being a point – so that part of the image will look sharp. The distance between the nearest and farthest plane that looks sharp is known as *the depth of field.*

The amount of depth of field that there is in a particular shot is not constant. It varies considerably. There are three main factors that determine how much depth of field there is in any shot:

- The focal length of the lens being used – the more wideangle the lens, the more depth of field.
- The size of the iris (or aperture) of the lens – the smaller the hole the light has to pass through the greater the depth of field.
- The distance at which the lens is focused – the closer to the lens the subject is, the less depth of field you have to play with.

Any one, or any combination, of these three factors can be used to control how much of the image is in focus. Fortunately, a camcorder allows you to see through its electronic viewfinder exactly what is in focus at that particular point. Unlike the viewfinder of a single lens reflex stills camera which normally only shows you the depth of field available if you were to use the maximum aperture available, a camcorder's viewfinder shows you the scene using the current iris setting – so you can see how much depth of field you have. Unfortunately, the small display does not make this easy – so you really have to look closely around the frame to be able to check what parts of the scene look out of focus.

Limiting depth of field

A classic example of when you might want to restrict how much of the image was in focus would be when you were videoing a person, but you did not want the viewer to be distracted by a cluttered background. For a stills photographer, throwing the background out of focus is easy, as all they need do is enlarge the aperture setting. Things are not as easy for the camcorder user. Not only do few camcorders have full manual iris adjustment, but by opening the aperture of the lens you need to increase the shutter speed of the camera to ensure the exposure remains the same. Increase the shutter speed too much on a video camera, however, and any movement in the frame will look jerky when played back. So using the high-speed shutter setting to restrict depth of field is only useful if your subject is completely stationary.

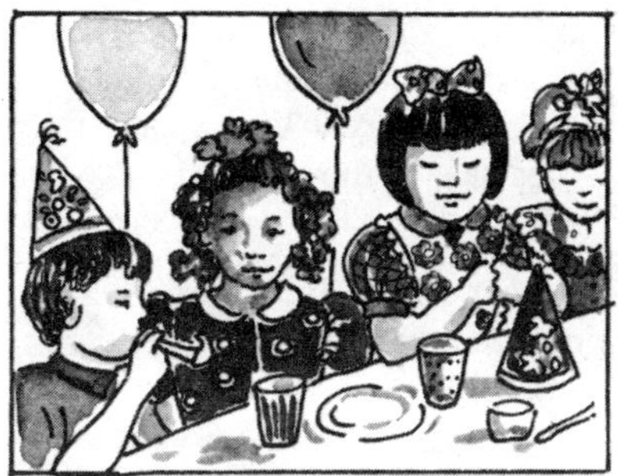

Extended depth of field – everyone in the room is sharply in focus

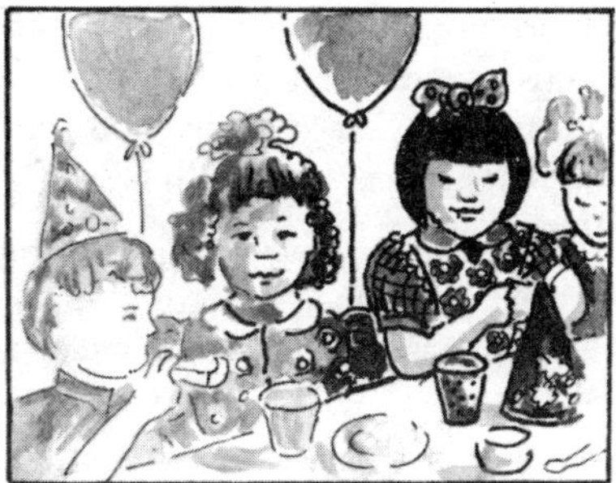

Limited depth of field – only one person is completely in focus

Restricting picture sharpness with a camcorder, therefore, frequently means making a compromise – using all three of the factors that affect depth of field. So to get the desired effect you will have to:

- Get as close to the subject that you can – trying to ensure that they are as far away as possible from the objects that you want to appear out of focus.
- Use a telephoto setting of your lens. Obviously, if you can get reasonably close to your subject (as above) then you will be no longer in a position to use a telephoto lens – but if you do have a choice between the two you will probably find that using a telephoto lens will give a better result, as the expanse of background will be restricted and less identifiable.
- Select a wider iris setting or a slightly faster shutter speed – remembering that the faster the shutter speed you select, the more juddery any movement in the frame will appear on playback.
- Another way to force the iris open automatically is to decrease the amount of light reaching the subject. This might not always be feasible. But if you can move the camera position or move your

subject so that they are not so brightly lit, this will force the auto-exposure system to compensate by opening the iris.

- Focus on a point slightly in front of your subject, so that the subject is still acceptably sharp – but the background becomes more blurred.

Increasing depth of field

In some situations you will want as much of the frame to be in sharp focus as possible. For example, if you are shooting a family gathering around the dinner table, you will want those nearest to the camera to be in focus, as well as those on the other side of the room. To achieve this, you will have to make one or more of the following adjustments:

- Use the widest setting feasible on your zoom lens.
- Increase the distance between you and your subjects.
- Ensure that you are not using one of the high shutter speed settings.
- If possible, increase the amount of light reaching your subjects to force the automatic iris to close down.
- Focus on a point between the closest and farthest subjects that you want in focus – so that both are covered by the depth of field available at this plane of focus. As a rough rule, this point of focus is approximately a third of the way between the closest and farthest points you want in focus.

Focusing

As depth of field can have such a great effect on what is and what is not sharp in a picture, it is absolutely essential that the camera is focused correctly. All camcorders have some form of autofocus system that makes this adjustment for you. No autofocus system, however, is absolutely perfect – and for this reason nearly all camcorders have some form of manual adjustment. Whichever way the lens is focused, however, it must be focused on the right part of the frame. While you know what the main subject in the frame is, your autofocus system does not – and can only guess using a set of pre-programmed rules. For instance, most autofocus systems will suppose that your subject is somewhere near the middle of the frame. If you then place your subject to the side, the autofocus system is likely to get it wrong. If there are two subjects towards the centre of the frame, the autofocus system is likely to assume that the closer of the two is the main subject. Although this is likely to be right most of the time, there may be situations when this is not so – and again the autofocus system will get it wrong.

Furthermore, autofocus systems will occasionally have problems focusing on anything at all, owing to the way in which they work.

Types of autofocus

Video cameras use two types of autofocus system – active and passive. The first measures the distance between camera and subject; the second analyses the picture itself and decides whether it is sharp or not. Both systems have their advantages and disadvantages.

Active autofocus systems measure subject distance by firing one or more beams of infrared light at the subject. By measuring the angle of the reflected ray it can work out the distance of the subject and move the lens accordingly. The farther the subject is from the camera, the narrower the reflected angle of the returning infrared beam.

The main advantages of infrared autofocus systems:

- It makes the necessary calculations quickly.
- It is very accurate for subjects near to the viewfinder (i.e. those distances where accurate focusing is more critical due to limited depth of field).
- It is not dependent on the contrast of the subject (so works well even if there is little contrast).
- It works well in low light (in fact, it will work in total darkness)

The main disadvantage of infrared autofocus is its limited range. It can only work up to a distance of about 10 metres. Beyond this it assumes that everything is so far away that it can effectively be judged as being the same distance. This distance is called infinity and is often referred to by the symbol ∞. Thanks to depth of field, with most lens settings, even with the widest aperture, this assumption is correct. However, with longer zooms (those with zoom ratios of 10x or more) this is not necessarily the case. With these, the difference between 10 metres and 15 metres is important when the light is dim (forcing the camera to select the widest iris setting available).

Other disadvantages are that it will not work through glass – as the infrared beams will tend to bounce back off the surface of the glass. So it is not very useful for shooting through windows.

It also will not work well with subjects that do not reflect the infrared beam (such as very dark, matt surfaces).

Passive autofocus systems use the image on the CCD imaging chip to detect sharpness. It works on the principle that there is more contrast in a sharp image than in an out-of-focus one. If no contrast is found, the camera will start focusing the lens until contrast is detected. The main advantages of the system are that:

- It works well at all subject distances.
- It works well with all zoom settings.
- It uses less battery power than infrared systems.

The disadvantages of a contrast-detecting autofocus system are:

- A tendency to be unstable, as it is constantly searching for a better focus.
- It needs a reasonable amount of light to work effectively.
- It won't work on low-contrast subjects (such as a plain-coloured wall).

Manual focus

Whatever autofocus system your camcorder uses, there are going to be times when it will not do exactly what you want it to do. Even so-called 'intelligent' autofocus systems, that use sophisticated programming to analyse the data to help minimise some of the problems outlined above, can still sometimes not work satisfactorily.

Unlike some mistakes that a camcorder can make, at least you can see when the autofocus is having problems. Either you will see in the viewfinder that your main subject is out of focus, or you will see, or hear, the autofocus system 'hunting' – that is focusing back and forth looking for the right setting. If this happens, you know it is time to switch to manual focus.

There are three different methods for manually focusing the lens commonly found on today's camcorders:

- A ring on the lens itself which you turn to bring the picture into focus.

- A thumbwheel control away from the lens itself which you turn to focus the lens.
- A pair of buttons – one to focus nearer and another to focus farther away.

Normally, to use each of these systems, you will first have to turn the autofocus off.

The easiest way to ensure that you have focused accurately on your subject is, before you start recording, to zoom in to your subject so that it fills the frame. Then focus manually. Finally zoom back so that the subject is the required size in the viewfinder. Because depth of field becomes more limited when you zoom in you will see that you have focused at the right distance much more clearly. As you zoom back you will still be focused at this distance.

Owing to the characteristics of depth of field, you will find that focusing is much more critical when the subject is very close to the lens. At these short distances, especially when using macro, you might find it easier to rock you and the camera back and forth slightly to find the right focus – rather than to run the risk of continually overshooting with the manual focus control itself.

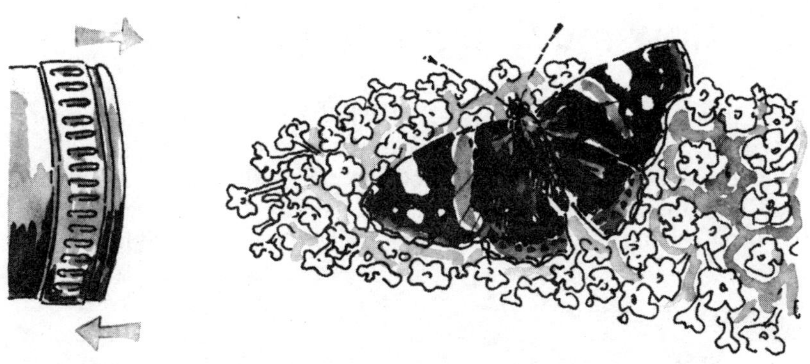

Macro focusing: often when videoing a small subject close to the lens, it is easier to move the camcorder back and forth to focus the image than to use the focusing controls on the camcorder

8
LIGHT

Just as we need light to see the world around us, so does a camcorder. However, there are important differences in the ways that the human eye and the camcorder's CCD chip see light. Our brains perform an extraordinary job in compensating for the very wide range of lighting conditions that our eyes are faced with every day. Although camcorders have sophisticated circuitry that tries to match the way in which our eyes see things, it has its limits. Knowing what your camcorder is going to see in a certain situation is part of the skill of an experienced camcorder user.

—— How much light do you need? ——

One area in which camcorders excel is in their ability to shoot in very low light. Manufacturers are always keen to tell you that their latest model can shoot in just three or four lux. What they do not tell you is how dark this really is. The truth is that five lux is equivalent to a dim, candlelit room. A well-lit living room at night would be 200 lux.

Lighting intensity

Dusk	5 lux
Candlelit room	5 lux
Street at night	50 lux
Well-lit room at night	200 lux
Sunrise/sunset	400 lux
Average daylight in winter	10,000 lux
Average daylight in summer	50,000 lux
Sunlit beach or ski slope	100,000 lux

On paper, all camcorders are easily capable of giving you a video image in even the lowest lighting conditions – even when there is not enough light to read by. However, this is not to say that camcorders give their best results in these conditions. Twenty lux will give you a much more tolerable picture than one shot at five lux – 100 lux would make it better still. Optimum picture quality is not possible until you have more than 1,000 lux available. Increasing the light ensures that the camcorder is not using its electronic gain system to boost the picture – and adding grain as a result. It also ensures that the colours of your subject are reproduced more accurately.

The most important thing to remember when shooting in low light is to ensure that you are using all the light that is available. Switch on all the room lights if you are indoors, for instance. Move your subject nearer the light, if this is possible.

Camcorders can have difficulties coping with very bright light as well. Beyond about 50,000 lux the picture can start to suffer again. On white sandy beaches or on snow-covered mountains, in particular, the intense reflection can prove a problem for the camcorder. In these circumstances it is worth fitting a neutral density filter to the lens. This filter is made of grey coloured glass or plastic, and cuts down the amount of light reaching the lens, without affecting its colour.

White balance

When we look at a white piece of paper we see it as white whether we are looking at it under candlelight or in the midday sun. In reality, different light sources have different colours. As bright sunlight is bluish in colour, our piece of paper will reflect this and should look slightly blue. A candle produces an orangy light, which again should make the piece of paper look slightly orange. However, our eyes automatically compensate for these changes, and we see the paper as white whatever.

The colour of a particular light source is known as its colour temperature and is measured in Kelvin (K). A normal household bulb (known as tungsten light, as its filament is usually made from tungsten metal) has a colour temperature of about 2,800K, while average daylight is about 5,600K. The colour of sunlight, however, varies greatly throughout the day – from the orange glow of sunrise to the deep blue of a summer sky.

Colour temperature scale

Candlelight	1,000K
Sunset/sunrise	2,000K
40 watt bulb (tungsten)	2,650K
150 watt bulb (tungsten)	2,900K
Quartz halogen video lamps	3,200K
Photoflood bulbs	3,400K
Hour before dusk/dawn	3,500K
Early morning/late afternoon sun	4,500K
Average daylight	5,600K
Lightly overcast sky	6,200K
Overcast sky	7,000K
Shade in summer	8,000K
Hazy sunlight	9,000K
Blue sky	20,000K

All camcorders have a system for making sure that colours are recorded as we would normally see them, whatever the light source – this is known as the auto white balance system. It gets round the problem of the different colours by electronically filtering the light so that it looks right. It is important that it chooses the right filtration for the situation – if it used a daylight filter for tungsten light for instance you would end up with a very orange picture. Using a tungsten filter in daylight, on the other hand, would give a cold blue appearance to all your footage.

The auto white balance system works well, but there are a few situations when it can work less than perfectly. Sometimes, for instance, it can work too well – when you are shooting a sunset, for instance. You want the camera to record the reds and oranges in all their glory – you do not want the white balance circuitry to filter them out. Some camcorders have ingenious systems for getting round this problem, while others have special program settings for you to use for sunsets. Alternatively, if you have manual white balance settings (see below) you should choose the daylight preset to give you the maximum colour for your sunsets.

Another situation in which some camcorders have problems is with fluorescent strip lighting. As this does not 'burn' constantly, like other forms of lighting, it does not fit neatly into the colour temperature scale. However, if you shoot under it with a daylight white balance setting you are likely to get an unpleasant green tinge to your shots. Some camcorders can correct this well, but others struggle. Be sure you know how well your machine can cope with it before shooting long sequences of video under it. If it cannot cope, use your own lighting instead.

The situation in which all camcorders will always have problems coping adequately is when you have two different light sources in the same shot. A common example might be a room indoors in early evening. The room lights are on, but there is still some daylight coming through the window. The problem is known as mixed lighting. There are two solutions:

The compromise. Make a decision as to which light source is the stronger. In the example above you could turn the room lights off and see if this has marked effect on the room brightness – if it does, the tungsten lighting is probably dominant, so set the white balance manually to tungsten. Alternatively, crop in closely for each shot, so that the effect of the two different light sources is not so apparent.

The professional solution. Avoid mixed lighting at all costs. Make do with the one light source if possible. If you cannot avoid it, filter one of the sources so that it matches the other. In the above example, you could put blue gels (sheets of coloured plastic) over the tungsten lights to make them like daylight. Alternatively, you could cover the window with an orangy gel to make the daylight appear the same colour temperature as the room lighting.

Because of the difficulties that an automatic white balance system can have in certain situations, some camcorders give you the option to set the white balance manually, when required. There are two ways in which this is done:

- By allowing you to choose from two or more different preset settings (tungsten, daylight, and sometimes fluorescent).
- Full manual control. This is set by placing something white in front of the lens under the particular lighting you want to use.

You can use a piece of paper for this, alternatively, such camcorders often come with a white translucent lens cap for this express purpose. With the white area filling the frame, you then press the white balance set button, and you are ready to shoot. The white balance remains the same until you change it.

Exposure

As explained in earlier chapters, a camcorder allows for the wide variety of different light settings it is faced with by altering a number of different controls. The main one of these is the iris – a circle of inter-leafing blades that forms a hole through which the light passes before it reaches the CCD imaging device. The bigger the hole, the more light that gets through.

Camcorders use the same f/stop system for measuring the size of the iris as is used in stills photography. The first place you are likely to come across f/numbers on a camcorder is on the side of your zoom

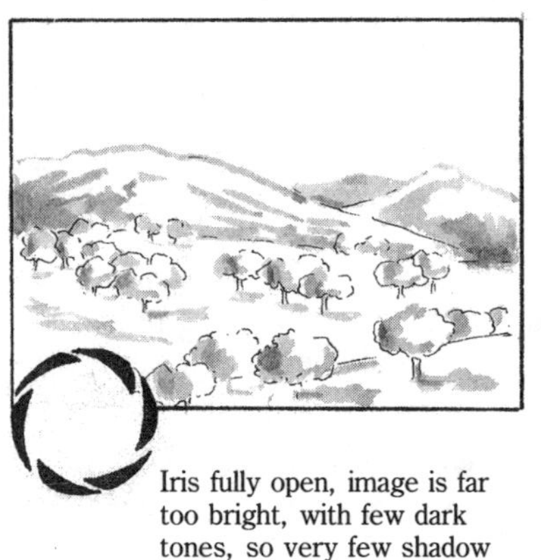

Iris fully open, image is far too bright, with few dark tones, so very few shadow areas

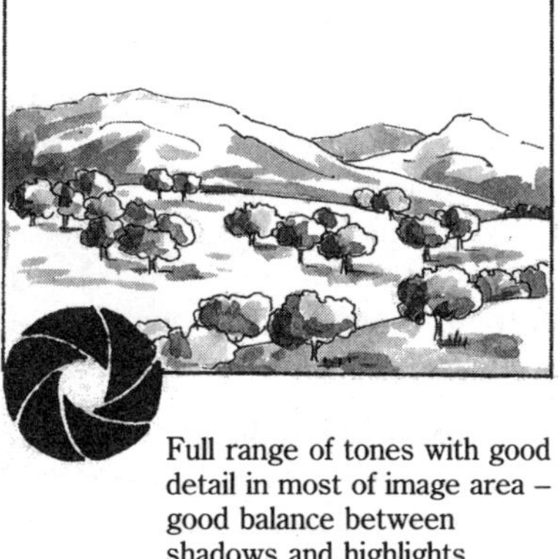

Full range of tones with good detail in most of image area – good balance between shadows and highlights

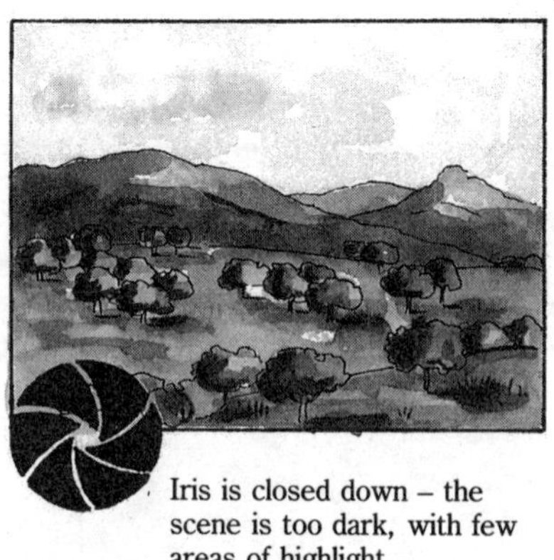

Iris is closed down – the scene is too dark, with few areas of highlight

lens. A set of three groups of figures is likely to be written here, saying something like:

6.5–52mm 1:1.8 Φ37

The first set of figures is the focal length range of the zoom (6.5–52mm). The third figure (Φ37) is the diameter of the filter ring in millimetres (in this case 37mm). The central figure (1:1.8) shows the maximum aperture of the lens (in this case f/1.8).

F/numbers are not a true measurement of the diameter of the hole of a particular iris setting. Instead, they show the relationship between the focal length and the diameter of the aperture. This works in such a way as that the focal length divided by the f/number equals the diameter of the aperture. Because of this, small f/numbers mean a large aperture, while large f/numbers mean a small aperture. More confusing still is the way that f/numbers increase. The simple way to remember it is that each time you halve the f/number, you are letting in four times as much light (e.g. changing from f/16 to f/8) – this change is also known as opening up two stops. Moving a 'stop' halves or doubles the picture brightness. The sequence of full stops goes as follows:

f/1   f/1.4   f/2   f/2.8   f/4   f/5.6   f/8   f/11   f/16   f/22

Very few camcorders actually display these aperture numbers when in use – so there's no need to learn this sequence.

As well as using the iris, the camcorder will also occasionally use two different adjustments to control exposure – the gain and the shutter speed. But as described in previous chapters this does mean some sacrifice in the picture quality. Increasing the gain means a grainier picture. Increasing the shutter speed means that any movement in the frame can look jerky.

Automatic exposure

All camcorders will take care of all the exposure adjustments necessary for you. Many, in fact, have little or no manual adjustment. This is due, partly, to the fact that auto exposure (AE) systems have improved enormously in the last few years. Typically today an exposure system will take brightness readings from twenty or more different areas across the frame. And by comparing these with a data bank of information stored in its memory, it can tell which type of lighting situation it has in front of it – and can make a good guess as to how bright it should look on the screen.

The main exposure problem that camcorders have is with high contrast subjects – that is where the brightest area of the screen is much brighter than the darkest area. The human eye can cope with areas with a contrast ratio of 100:1 – where the highlights are 100x brighter than the darkest shadows. Camcorders, however, can only cope with a contrast ratio of about 30:1. Some scenes that it may have to cope with, such as a sunny beach, might have a contrast ratio of 500:1. In such high contrast situations, the camcorder cannot expose for every part of the scene correctly. Some of the highlights will be burnt out to white, or some of the shadow areas will become completely black – or both. The camcorder, therefore, has to make a compromise – ideally one in which the parts of the picture that you really want the viewer to see are exposed correctly.

The most common example of a high-contrast situation is when your subject is standing by a window. Even the best auto exposure system is likely to be fooled by the enormous amount of light around the edge of the frame – the light coming through the window. Because of this, it closes down the iris, and your subject becomes a black silhouette. Part of the picture, however, is correctly exposed – the window itself, as you can still see what is through it on the video.

To compensate for this common situation, most camcorders have a backlight button. This will open up the iris by a preset amount (usually a couple of stops). The subject standing in front of the window is then better exposed. However, the window itself is now overexposed – so you cannot see what is through it.

Sometimes with high-contrast subjects it is not obvious which of a whole range of exposures is the correct one. For instance, if you are shooting the city lights at night, there is a great deal of difference in the brightness between bright white lights and the darkest shadows. But although the lights will look brighter if you open the aperture, the black areas will not change significantly. So you do not have to worry too much about exposure problems – you are likely to get a presentable picture whatever. The key thing to watch out for here (if your camcorder allows you to do anything about it) is that the automatic gain does not try to boost the picture brightness. The gain will add unnecessary noise (grain) to your picture without doing anything to improve the shot. Some camcorders have a special program exposure mode to deal with this situation.

Other situations where your automatic exposure system can be fooled is when the frame is filled with predominantly black or white subjects

– a white cat on a white kitchen floor, or someone dressed in black against a black wall. In these situations, even though there is little contrast, the auto exposure system is likely to try to even out the brightness – so that in both situations you get a greyish picture, rather than one that is all-black or all-white.

Fortunately, with all these exposure problems, the electronic view-finder at least provides you with a rough guide to when things are going wrong. It shows you the exposure as it is currently set, so you have an indication of when to use special program settings, or full manual iris control. Even if your camcorder gives you none of these overrides, at least you know that the shot you are attempting is unlikely to work – so that you can change the lighting or your camera angle so that it does.

Quality of light

There is much more to lighting than getting the exposure right. Although you can video subjects in a wide variety of light intensities, the results do not always look the same and it does not necessarily follow that bright sunlight gives the most attractive-looking results. A castle videoed in the midday sun can look flat and colourless, but caught by a single ray of sunlight just after the storm it can look golden and majestic.

The way we see things is affected by the angle at which the light falls on them. To use an extreme example, if the a subject is lit from behind it can look a black mass – a silhouette. Lit from the front, we see it in all its true colours. You must remember that the way we, and the camcorder, sees a particular subject is by the way that the light is reflected off it. Through this process we get all the information about the subject's shape, colour and texture – but the amount of detail we get about each of these is dependent on the angle of the lighting.

The common types of natural lighting you will come across while videoing can be broken up into four different categories:

Frontal lighting. It was only a few decades ago that photographers were warned to make sure that the sun was directly behind them when taking their pictures – this is known as frontal lighting. It still has great advantages for the modern videographer. There are hardly any shadows to hide areas of the picture – as they mostly fall behind the subject, where they are obscured from view. This means there is less contrast in the picture than with most other forms of lighting – so there are fewer problems getting the exposure right. On top of this, colours are seen at their maximum brilliance – and for this reason alone, frontal lighting is often the best way to shoot a subject.

There are drawbacks, however, with having the lighting directly behind the camera. Without any shadows, subjects can look flat and two-dimensional – like cardboard cutouts. Also, the lack of shadows means that you can tell little about the texture of your subject – you can see the colour of an orange in all its glory, but you cannot see its dimpled skin. There are other pitfalls – such as people screwing up their eyes as they look towards the camera, or the shadow of the camera operator appearing in shot.

frontal lighting

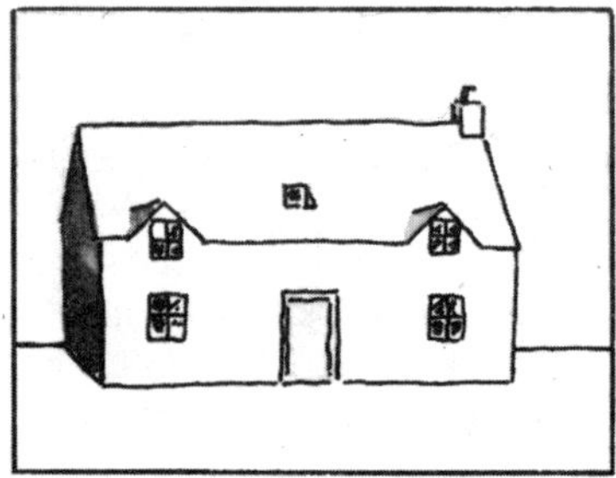

Shadows fall behind the subject. Good colour saturation and detail, but no feeling of depth

backlighting

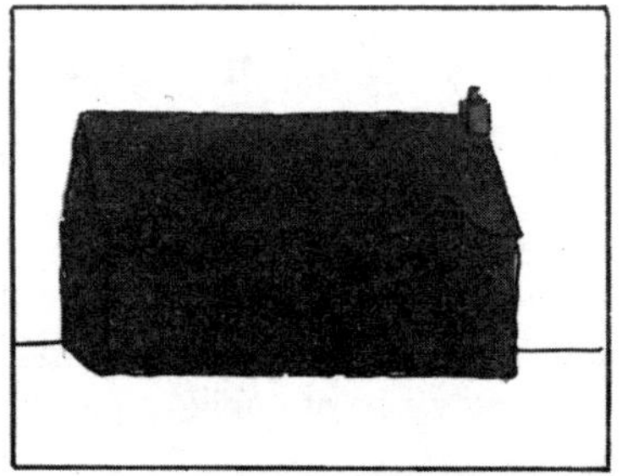

Silhouette shows outline shape only, and foreground is obscured by further shadows

rim lighting

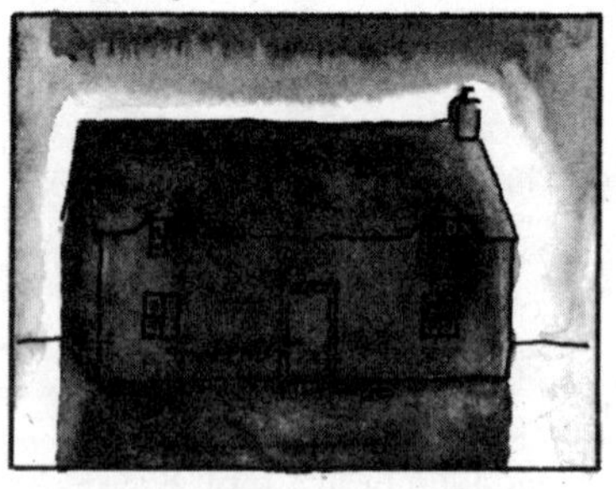

If subject is lit from directly behind, it can give a rimlit effect – emphasising its shape against a dark background

sidelighting

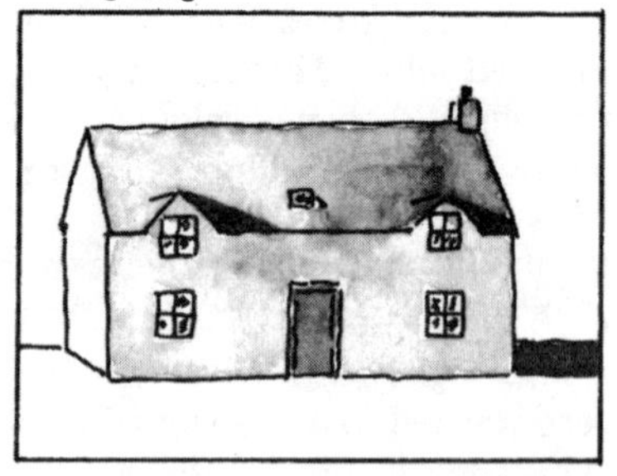

Sidelighting gives a strong feeling of depth, but detail is often obscured by the harsh contrast between the lit areas and those in shadow

However, frontal lighting is the safest way of taking most shots – giving picture postcard lighting that is sure to please. And if you put the sun just slightly to the side – so it is over your shoulder, rather than behind your head, it will give a slight bit of shadow to the side of your subject, helping to make it look that touch more three-dimensional.

Backlighting. Having the sun directly behind your subject means that little light is reaching the side of the subject you are videoing. It therefore appears as a dark mass – a silhouette. Little or no information is available about colour or texture. All you have is a bold outline. This can suit some subjects well – whose shape is bold and interesting enough. Statues are a good example – or a person doing something that is instantly recognisable from the outline, such as an archer firing a bow and seen side on. In fact, there is always more information available than it at first seems – as the movement suggests more about the subject's identity.

There are two specific cases when backlighting has a special use when videoing; these are *rim lighting* and *translucent lighting*. Rim lighting occurs when the subject is backlit, but the edges or the subject reflect some light directly at the camera. It is most obvious when the background is also dark. It works particularly well in misty or dusty conditions, when the particles in the air around the subject reflect the light – making it look like there is a halo around the subject.

Translucent materials, for obvious reasons, often look at their best when lit from behind. As the light does not pass directly through them, but is dispersed in all directions, the light is much softer – showing up the colour and the internal patterns of the material to their full. Leaves, flower petals, crystals, stained glass are all good examples of materials that suit backlighting well because of their translucency.

Sidelighting. When an object is lit from the side, it becomes a mass of highlights and shadows. Protruding parts of the subject and other parts that are facing in the right direction are caught in the light – and are seen in full colour. Other parts of the subject that are facing away from the light or hidden by the protruding parts become bathed in deep shadow – so show no colour at all. Most interestingly, there are middle areas facing towards the camera that are neither highlighted nor in shadow – here the texture of the subject is greatly exaggerated. With simple-shaped subjects, sidelighting works well – there is enough detail about colour and texture in some areas to give an idea of the whole, and the shadows give an exaggerated feeling of three-dimensionality.

However, with more complicated shapes the frame can become a confusion of shadows, where nothing is clearly visible.

Moving subjects can work well with sidelighting, as the shadows will continually change – even with small turns of a person's head, for instance. Detail which may be hidden one second will momentarily be lit up and identified for the viewer the next.

Diffuse lighting. In direct sunlight the shadows cast by any subject are distinct – there is a clear division between light and shade. This is known as hard lighting. However, sunlight is not always direct. Sometimes it is diffused by cloud cover, or has been reflected off other surfaces even before it reaches the object you wish to video. Shaded areas on a summer's day are not directly lit – but they are not completely dark: light reaches this area by reflections off neighbouring buildings, or by being scattered by particles in the sky.

Diffuse lighting casts very little shadow, as nearly all sides of the subject are equally illuminated. You therefore get a lot of detail in the picture, but little information about texture or depth. However, the lack of heavy shadows means that there is no high contrast for the camcorder to contend with. It also gives you the freedom to shoot your subject from almost any angle, without worrying about how the lighting will change as you alter your viewpoint.

Using video lights

A video light is a handy, relatively inexpensive, accessory which can come to your rescue when lighting levels get low – especially when videoing indoors at night. All a video light does, however, is to increase the quantity of light – many on the market do little to improve the quality of the light.

Many video lights that you can buy, for instance, are designed to be fixed to the camcorder itself – either via an accessory shoe, or by a special bracket that attaches to the tripod bush or the battery mount. The disadvantage of using a lamp in this position is that it provides a hard directional light that is likely to cause a hard shadow on the wall immediately behind the subject you are videoing. If there is no wall behind, there is still a problem. Video lights only work over a short distance – as the distance between the subject and the light doubles, the amount of light reaching the subject is reduced by 75% (a property known as the *inverse square law*). So if there is nothing behind the subject (or even if it is more than a couple of yards behind) it will appear as darkness on the video and this can look rather artificial, especially when shooting indoors.

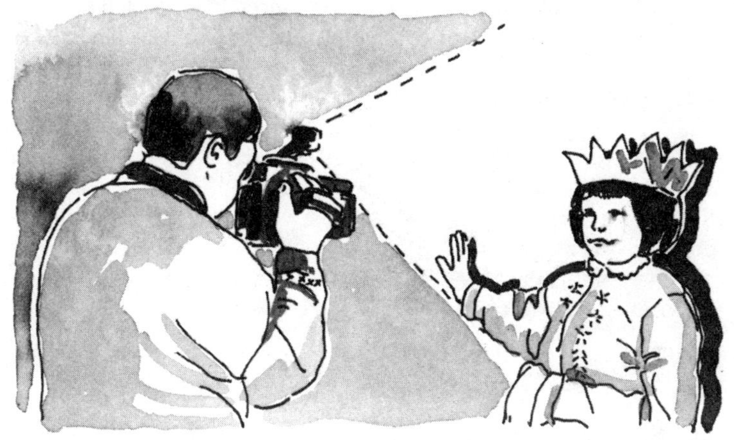

Having a single video light on the camcorder means, although the subject is better lit, a harsh shadow is cast on the background

Video lights can also kill the atmosphere of a room that is meant to look dark or shadowy. A great example of when a video light does more harm than good is at a birthday party when the cake arrives – the scene should look like it's lit by candlelight, not like it's be illuminated by a giant floodlight.

It is, of course, possible to recreate any lighting condition using artificial lights. The professional movie business does this all the time, turning day to night, and night to day to suit their tight schedules. For the camcorder user working alone this is rarely practicable – unless there is plenty of time to prepare beforehand.

A good tip therefore is to try to make do without a video light whenever possible. Turn on all the normal lights in a room – desk lamps, reading lamps, anything you can muster. Change to higher wattage bulbs if necessary. Move your subject, if possible, to a part of the room that is better lit. Once you have done all this, you can then, if necessary, use the video light. As the light level in the room has been increased it will have a lesser effect, and its shadows will be less noticeable. If possible, hold the light well to the side of the camera and as high as you can.

If you can make none of these alterations to the existing lighting in a room, you can try bouncing the light off a nearby wall or ceiling. This will soften the lighting so that you will avoid any hard shadows. This works best with white walls, not only because they reflect more light, but because coloured walls will reflect coloured light, which might make your subject look slightly bizarre. Bouncing the light is really only possible with relatively high wattage lamps – those rated 100W or larger.

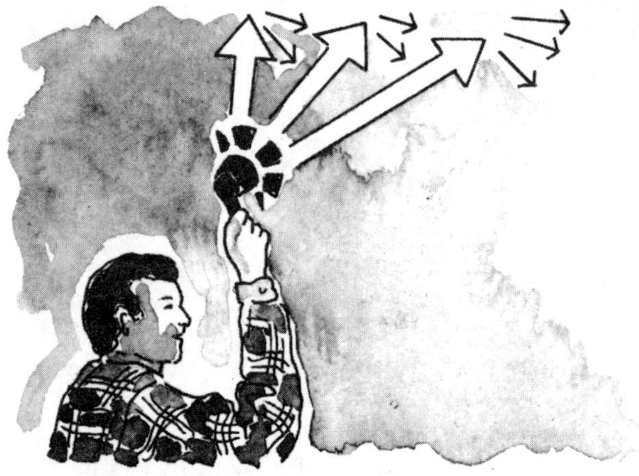

Bouncing a video light off a ceiling turns it into a diffuse light source – giving fewer problems with harsh shadows

The key thing to remember when using artificial light is to pay attention to the quality of the light – not the quantity. As described at the beginning of this chapter, it is possible to shoot video in the dimmest of light – don't add extra lights unless it is really going to make the picture look better.

9
FRAMING YOUR SHOTS

A 'shot' is the basic building block of any video. It lasts from when you press the button to start recording until you press this button again to put the camcorder into pause. It is very tempting when you start to use video to make these shots last for minutes at a time. However, if you watch any television programme or movie you will see that each shot lasts only a few seconds. It is this technique that is fundamental to good video-making and one that you must learn to do automatically. This continual stop-start process might seem disjointed at first, but it allows you to change camera angles and shot size between each shots. This adds visual variety to your videos and this is what makes them interesting to the viewer.

In the days of cine, there was a major incentive to keeping shots as short as possible. Film was expensive and each roll lasted only a few minutes. Wasting film was a costly business. Video does not give you such a strong incentive. The tapes are relatively cheap and can last three hours or more without the need for replacing. What is more, you can re-record over tapes that do not quite hit the mark. All the same, this is no excuse for making boring movies. Through television, we have all become used to a certain standard of video production. This standard demands the stop-start method of shot variation and your viewers will expect it if they are going to be entertained by your videos.

Calling the shots

By altering the focal length of your zoom or by moving backwards and forwards, it is possible to change the size of your subject in the frame.

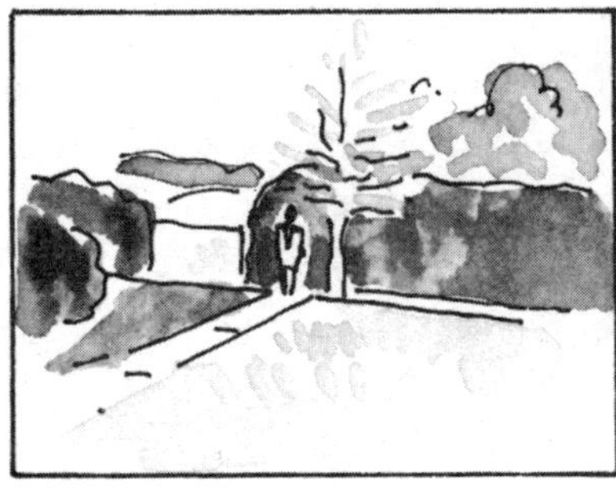

ELS – extreme long shot. Wide shot where person fills just a small part of the frame

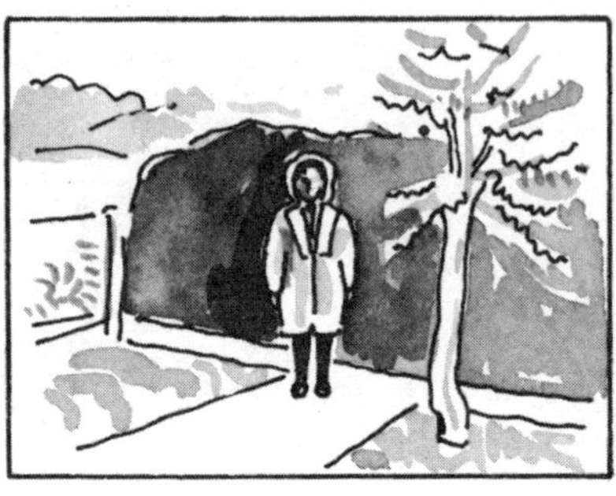

LS – long shot. Person fills half to two-thirds of the screen height

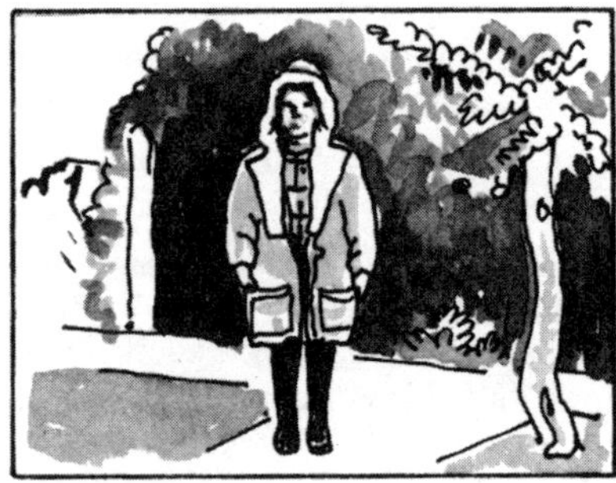

MLS – medium long shot. Person fills frame height

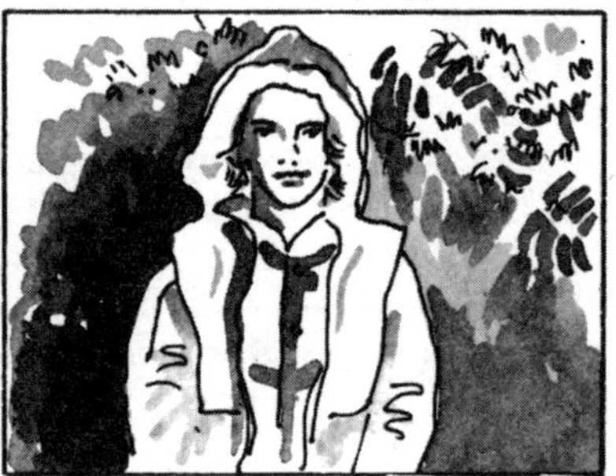

MS – mid-shot. From waist upwards

The different magnifications in which a subject can be shown on the TV each have their own names. Generally these are described in relation to how much of a person appears in the frame at one time. The three most common shot types are *the long shot* – a full figure shot; the *mid-shot* – from the head down to the waist; and *the close-up* – a head-and-shoulders shot. However, there is a whole range of shots in between and beyond these which can be used for specific purposes. Here is the full list of what the shots are called, and what they are used for:

Extreme long shot (ELS) – This shot is used for shooting a wide panorama or a whole building. Not usually used for shooting people, as they appear as small specks on the screen. However, it does give a good idea of the general surroundings – if it is important to use these.

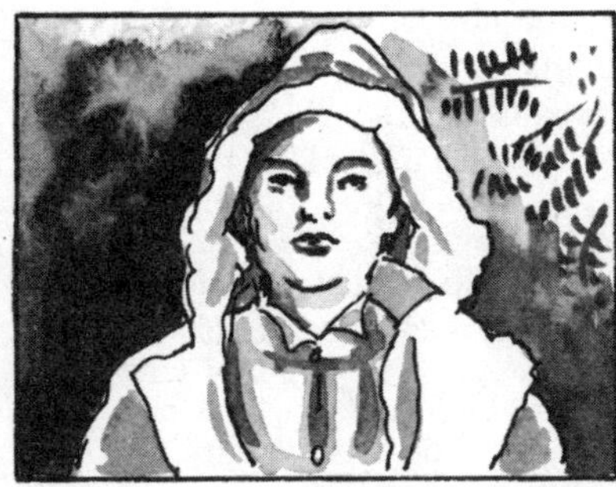

MCU – medium close-up. From chest upwards

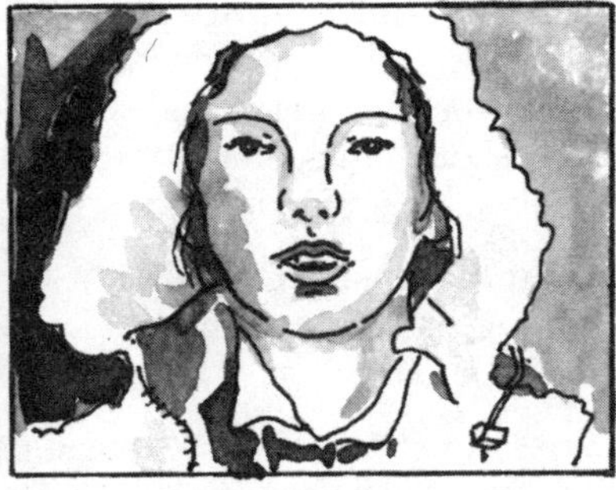

CU – close-up. A head-and-shoulders shot

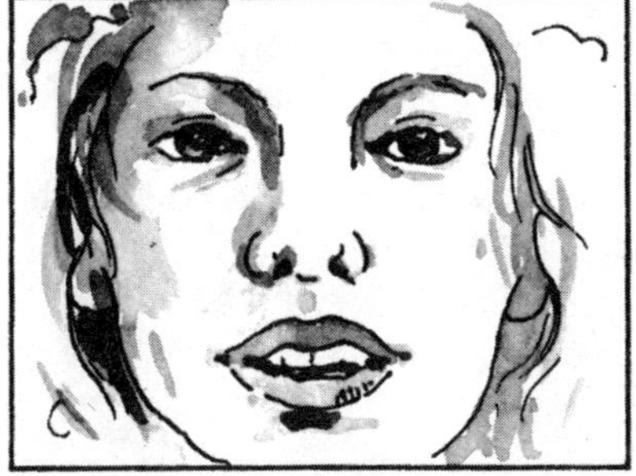

BCU – big close-up. Full face shot

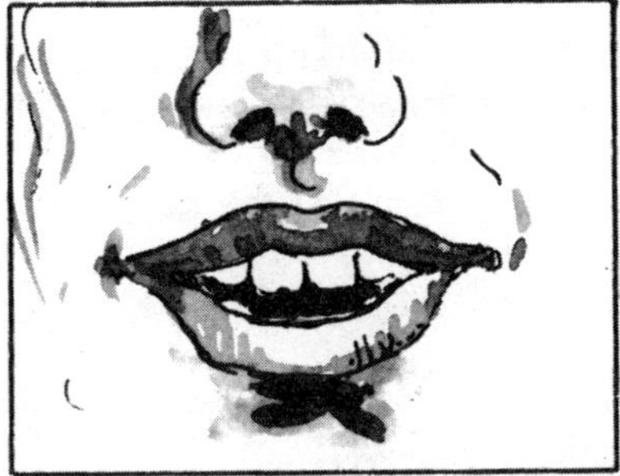

ECU – extreme close-up. Fills the frame with part of the body

It can also be used as an establishing shot, one of the first shots used in any scene, to show the viewer exactly where the sequence is being shot.

Long shot (LS) – One in which a human figure occupies between half to two-thirds of the height of the screen. One of the shots that you should use most often. It shows you exactly what a person is doing, where and with whom. Not so useful if you want to hear what the person is saying. It can also be used as an establishing shot.

Medium long shot (MLS) – One in which a person occupies practically the whole screen height. However, keep their head and feet slightly within the frame, otherwise the shot will look awkward. Good for showing groups of people.

Medium shot or **mid-shot (MS)** – One showing a person from just below the waist upwards, allowing a small amount of space above the head. Here the person is easily identifiable on the TV screen, and is close enough to see facial expressions. It can be used for two or three people talking together.

Medium close-up (MCU) – Cut from the breast upwards, again allowing space above the head. Good for showing two people having a conversation.

Close-up (CU) – The head and shoulders shot. Cuts from the top of the arms to just above the head. The ideal shot for showing one person's expressions as they are talking – not too close, but close enough.

Big close-up (BCU) – A more dramatic way of portraying a single person, so the head neatly fills the frame – from chin to top of the hair.

Extreme close-up (ECU) – Should be used with extreme caution. Fills the frame with just part of the face – and is designed either to shock the viewer, or to show some small detail which would otherwise be unnoticeable in the frame. At a wedding, for instance, you might use this shot to show the ring on the bride's finger.

Shot length

'How long should each shot last in my video?' The simple answer to this $64,000 question is 'As short as possible'. At the end of the day, a 30-minute movie is likely to more enjoyable for the viewer (even if it is a member of your family) than one that drags on for an hour-and-a-half. The shot is the basic building block of your video – if you can make these short, you can cram in more information in a much shorter time.

The basic rule of shot duration is that long shots should last longer than close-ups. A long shot has lots of different things to look at and it takes a while for the viewer to assimilate exactly what is in the frame. A close-up, however, has just one subject which is instantly recognisable, so it only needs to be on screen for the briefest amount of time.

As a guideline, you would expect a long shot, especially one that shows you a new location (an establishing shot) to last ten seconds or more. A close-up – especially if it is a close-up of something we have already seen in a wider shot – need only be recorded for two or three seconds.

Mid-shots should last somewhere in between – again depending on how much new information there is for the viewer to take in.

If there is any movement in the shot, it should be recorded for slightly longer – as the viewer will want to watch what's moving – so add about another 50% to the above timings.

If there is something in the shot that the viewer has to read – a sign, for instance – the shot should last as long as it should take to read it. A rough guide here is to allow a second for every two words.

Sometimes you will find that the subject matter itself will dictate the pace at which you stop and start recording. A bustling city might demand a succession of short shots to show its busy pace. A country village might need longer shots to help convey its tranquillity.

Most important of all, your soundtrack will often dictate when it is safe to stop recording. You do not want to cut someone off in mid-sentence – not just because you will miss what they were going to say, but because it will sound awkward when you replay the tape.

Shot composition

There are no hard and fast rules about how you should arrange your subjects in the frame, but there are plenty of guidelines which can be of help to those that who want them. Many of these have been passed down through the history of art – from painting, through photography and film, and finally to video. Video, however, is perhaps one of the most limiting mediums in terms of composition – as every shot has the same proportions. Every frame is four units wide to three units high – the shape of a standard TV set. Upright shots are not possible.

Although video is about recording movement, many of the individual shots that you will shoot will have little or no movement in them. They can, therefore, be composed as stills. Even when there is movement in a single shot, you will still have some options about how to frame the shot – even if its composition will change during recording.

No-go areas So they can show the biggest image possible, most TV sets are usually adjusted so that the edges of picture are not seen. Your camcorder's viewfinder, however, tends to show the whole frame. For this reason you should always avoid putting important parts of the picture at the extreme edges of the frame – just in case they are not shown when you replay the video through a television. Titles, in particular, must be positioned well away from the sides for this reason – so they retain some space around them when they are shown.

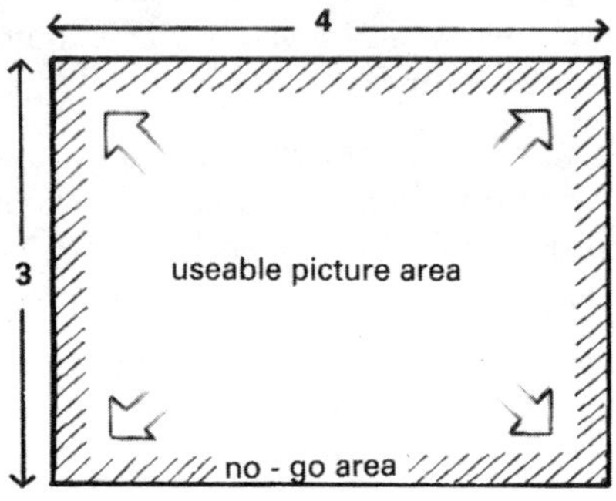

Body shots We saw above that one of the most important things to remember when videoing people is to leave some headroom – that is a small amount of space between the top of the frame and the top of the person's hair. This should be done with all shots except those big close-ups and extreme close-ups. The other thing that you should avoid is to cut the shot off through the joints in the limbs – particularly through the legs or the hands. The same applies to the shoulders, elbows and knees – either leave these in the shot completely, or cut them out entirely.

Looking space When we look at a picture of a person's face the first thing we normally look at is their eyes. If the person is not looking towards us, our natural tendency is to follow the direction of their gaze. It is therefore a convention when framing a shot of a person looking off-screen – whether they are in profile or three-quarters-on to the camera – that you leave more space in front of them than behind. This is known as 'looking space'. Using it reinforces the sense of direction inherent in this sort of shot. Not leaving this space – or leaving it on the other side of the frame – makes the shot look extremely uncomfortable, almost as if they are looking at a wall from a couple of inches away.

Moving space Just as you should leave space in the frame for people to 'look' into, you should also leave space for them to 'move' into when they are moving across the frame. Whether it is a person walking or a car at top speed the same principle applies. If they are moving towards the right, place them at the left side of the frame. If they are moving to the left, place them on the righthand side of the frame. Again this reinforces the feeling of movement in the shot – and stops the picture

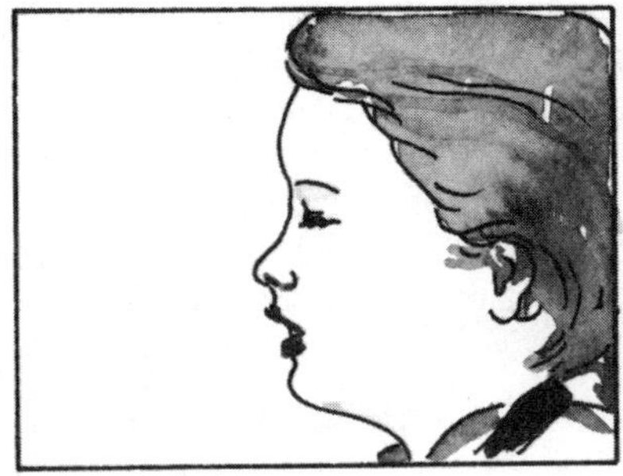

Adequate looking space

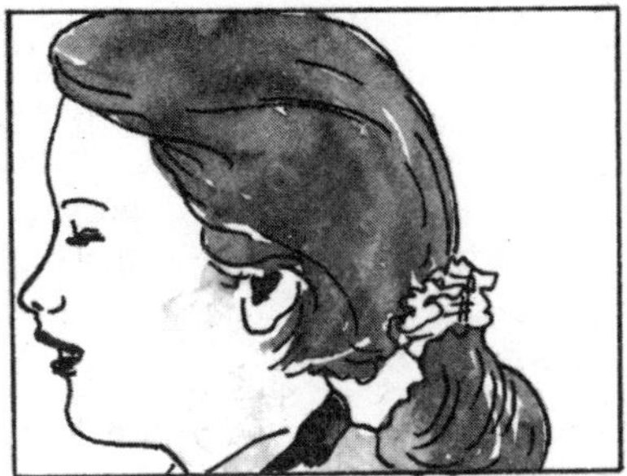

Inadequate looking space – this shot looks uncomfortable

looking uncomfortably cramped. We will see how you can follow this type of movement, retaining this moving space, in the next chapter when we look at panning and tracking.

The rule of thirds One of the oldest ways of composing a picture is to use the rule of thirds. The basic principle here is that a picture in which the main subject, or part of a subject, is placed slightly to the side of the frame will look more appealing than one where the subject is dead-centre. The way the rule of thirds works is that mentally you divide the frame equally into thirds both horizontally and vertically. This gives you a grid with four lines – and four points where these lines intersect.

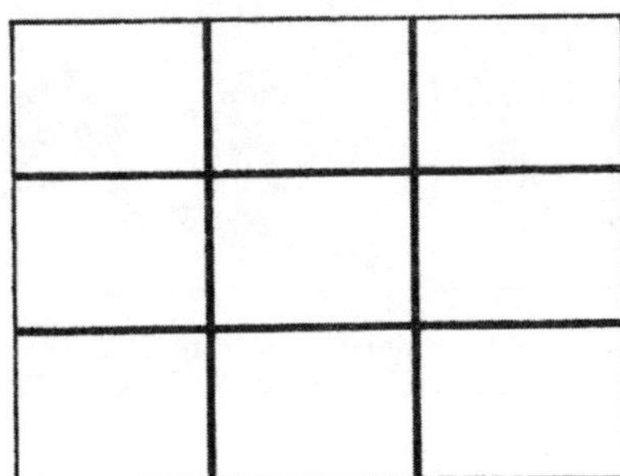

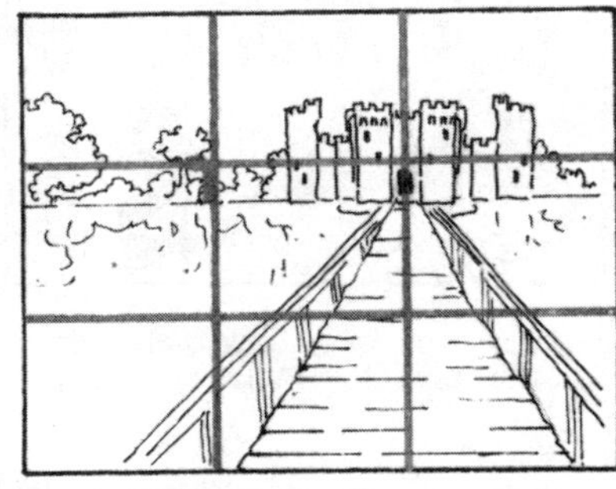

Rule of thirds: shots should be composed so that subjects should fall on the intersection of the four lines, or on the four lines themselves. Here the main subject (the castle) falls on one of the intersections, and the horizon falls on one of the horizontal lines of the grid

To compose your shots, you should aim to get your main picture element on one of these intersections – or on one of the lines themselves. For instance, if you were shooting a wide shot of a distant castle on a hill, you might place the castle on the top right intersection. But at the same time the horizon (the break in the picture between sky and land) would be on the top horizontal line. With a big close-up of someone's face, you may be able to place eyes over the top two intersection points.

In practice, it might be hard to visualise this grid of lines over your viewfinder. For most people, it is a lot easier just to remember to put the most important part of the shot slightly off-centre.

Dynamic diagonals Videoing things from straight on can make them look flat and lifeless. If you take a shot of a building, for instance, standing square in front of it, all the lines of the building will be parallel to one of the two sides of the frame. To make a shot look more dynamic, it is better to compose the shot so that includes a number of diagonal lines and preferably ones that are not parallel to each other. One of the best ways to do this is to get closer to the subject and use the wideangle setting on your zoom. In the example of the building, for instance, if you stand to one side near one of the corners of the building with a wideangle lens – the horizontal lines of the building will become diagonals.

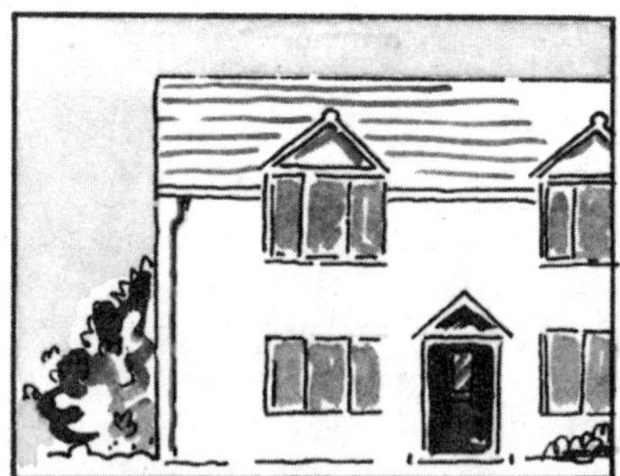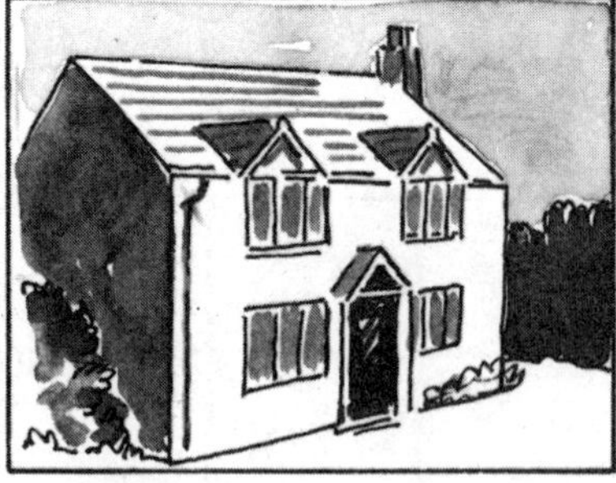

A change of viewpoint converts uninteresting parallel lines into dynamic diagonals. Here a distant telephoto shot has been changed for a closer, tilted, more wideangle shot

Similarly, if you have two items to frame in a particular shot, they often look better if the imaginary line between them makes a diagonal – not least because it makes better use of the total picture area, than placing them in a line that is parallel to the sides of the frame.

Dynamic diagonals can be over used, however. They are meant to look different, but if used all the time will look the norm. Also, some shots will look decidedly peculiar if you put things on the slant. Horizons, for instance, should be rigidly kept on the level.

Lead-in lines Another use for diagonal lines is to lead the viewer to the main subject in the picture. The classic example here is the landscape shot – where a road or river leads your eyes through the shot to the horizon. Because of the properties of perspective, the parallel edges of the road seem to converge – forming an arrow which seems to say to the eye 'Follow me'. Just as useful on this occasion are curved lines, which perform the same task of leading you through the shot, but in a more subtle fashion.

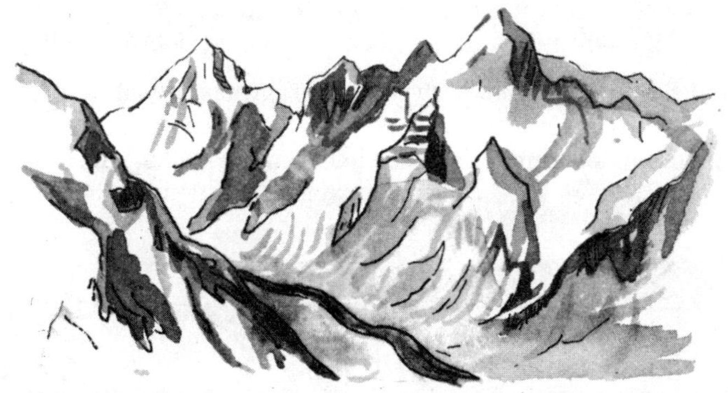

Lead-in lines: the river leads your eyes through this mountain range

Filling the foreground One of the main problems with wide landscape shots is that the area of interest in the shot is a narrow band at the top of the picture. There is little of interest in the foreground of the picture. One way round this problem is to frame the shot so that the horizon falls a third of the way up the frame – this way you include much more sky, which may be interesting in its own right, and there is no real foreground to the picture. However, if the sky is much brighter than the rest of the picture, or if it has no colour or clouds to add interest, this approach will not work.

Another way around the lack of foreground interest is to find something that will fill this empty space. A brightly coloured flower, such as a poppy, an interestingly shaped rock, or a farm gate can be used to fill this otherwise empty space. All you have to do is to get close enough

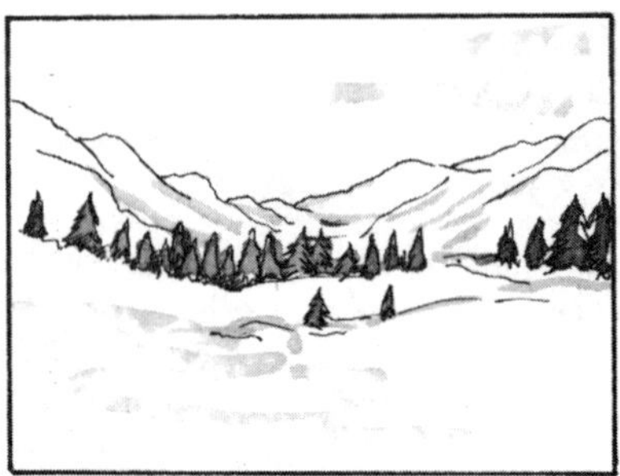

Nice landscape – but empty foreground means wasted, bland, space in frame

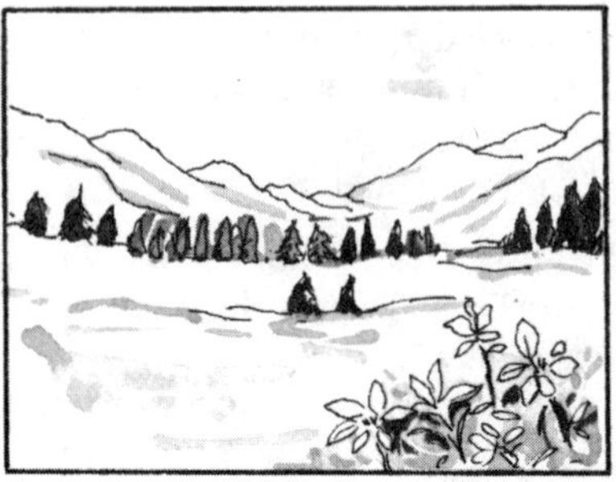

A slight repositioning of the camera finds flowers to help fill the foreground

to this object so that it fills a disproportionate part of the frame. This has the added advantage that it adds a feeling of depth to the shot.

Natural frames Another way to add foreground detail to a shot is to use the frame-within-a-frame approach. Here a natural frame is used to form a window through which the camera looks. Archways, the branches of a tree, and, of course, window frames can all be used in this way.

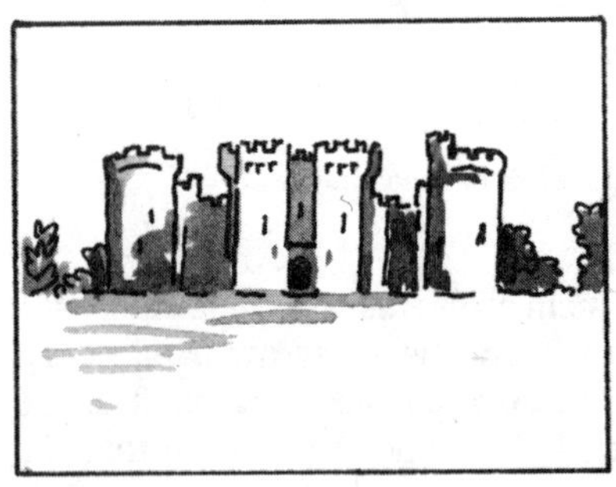

An empty foreground – you could get rid of it by zooming in closer . . .

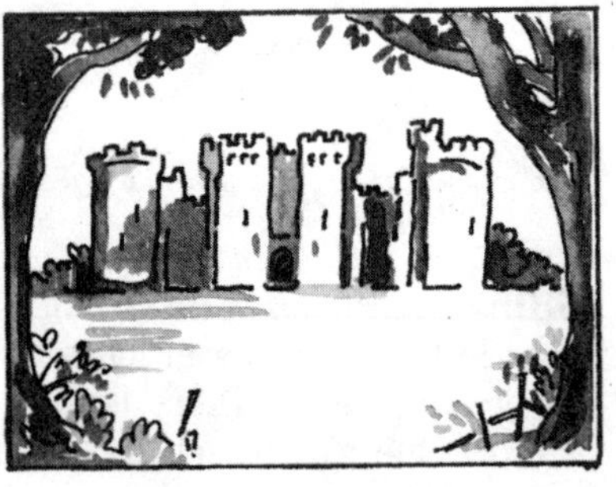

. . . however, by repositioning the camera slightly, you could use trees to make an archway or natural frame for your subject

Camera angles

Most of the time, you will be using your camcorder at head height, with the lens roughly parallel with the ground. However, there are

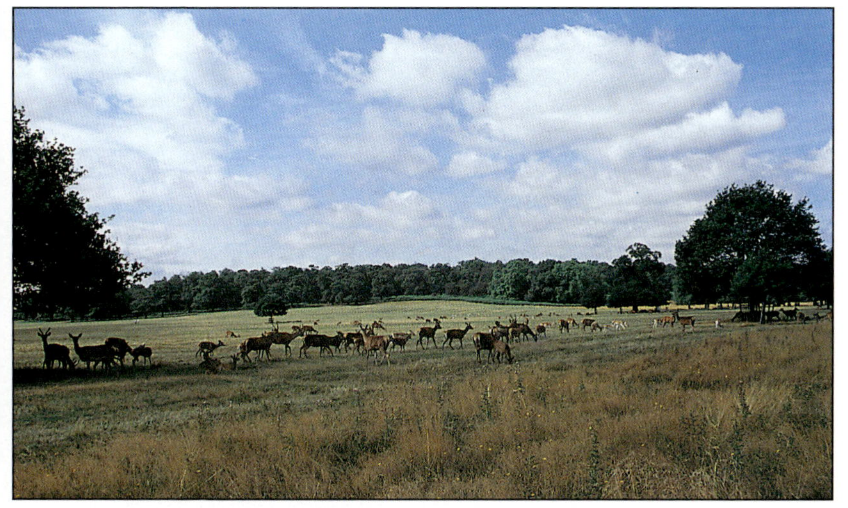

▲ **Figure 1** Maximum wide-angle zoom setting; this should approximate to a 35mm lens on an SLR stills camera

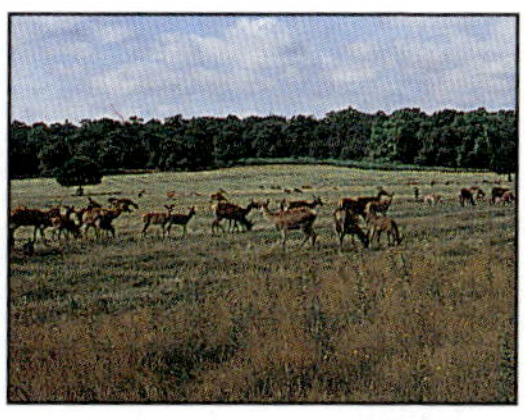

▲ **Figure 2** 2x zoom setting

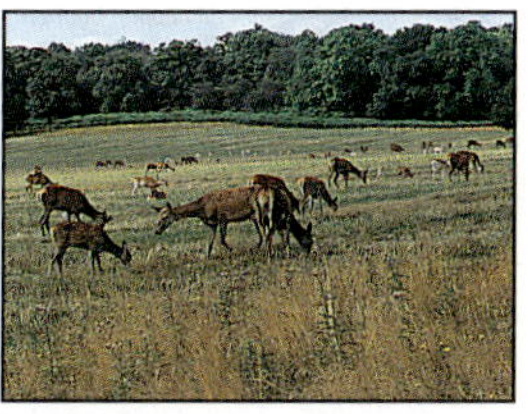

▲ **Figure 3** 4x zoom setting

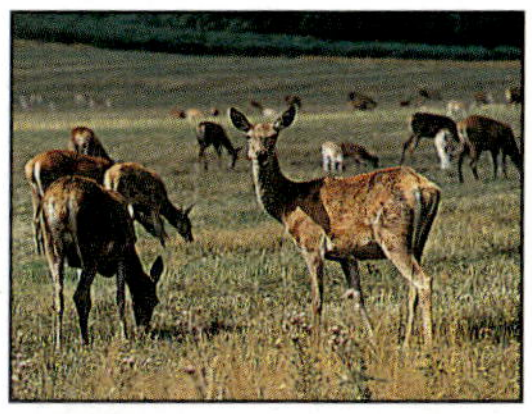

▲ **Figure 4** 8x zoom setting

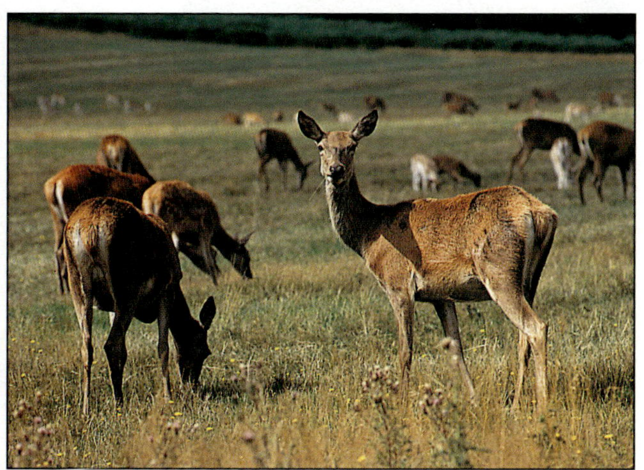

◄ **Figure 5** Maximum telephoto lens setting possible with a camcorder with a 12x zoom

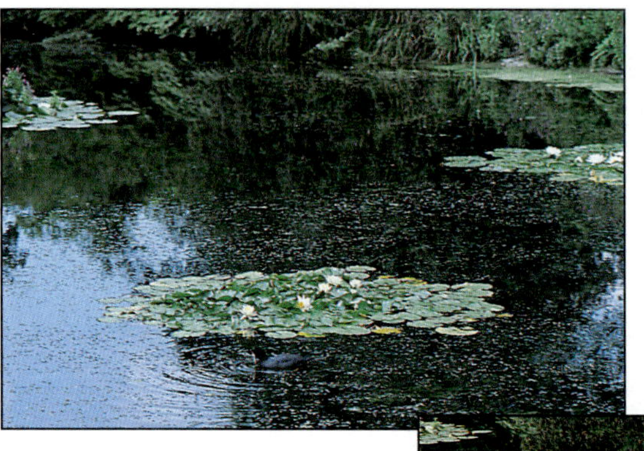

Figures 6 and 7
The white balance is set automatically by all camcorders. Only some allow manual adjustment. In the picture on the left, the blue colour cast is caused by using a tungsten white balance setting in daylight. The correct daylight setting is used in the shot below

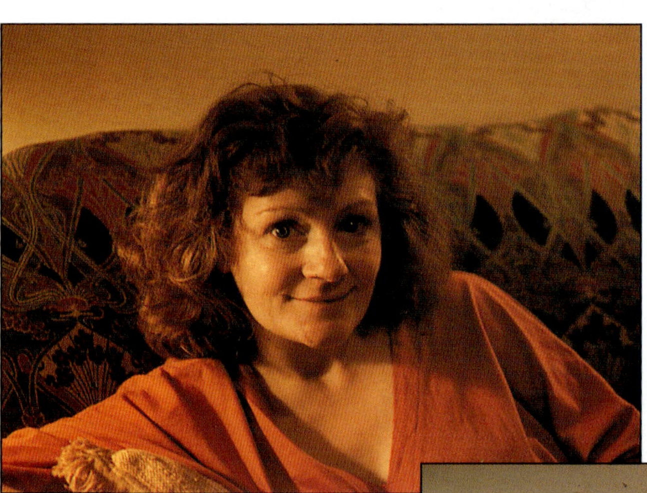

Figures 8 and 9 Tungsten lighting from ordinary household bulbs gives an orangy picture if used with the daylight white balance setting (above). Using the tungsten setting (right) gives more natural colours

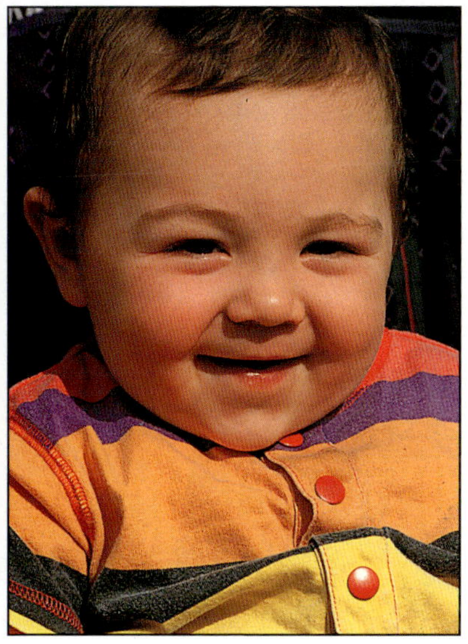

▲ **Figure 10** Frontal lighting, with the sun behind you, gives you the best colours

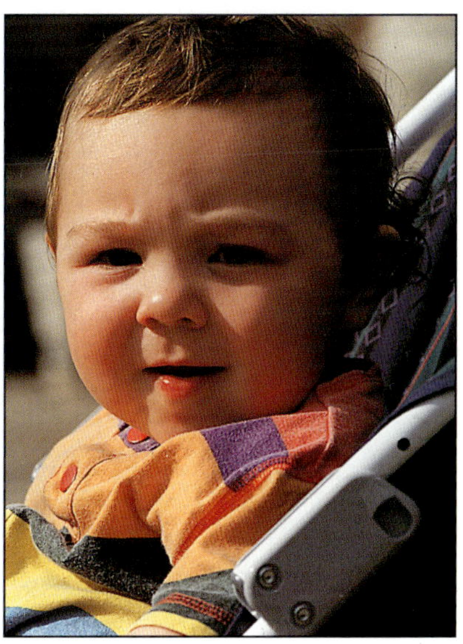

▲ **Figure 11** Sidelighting emphasises form and texture, but some detail is lost in shadow areas

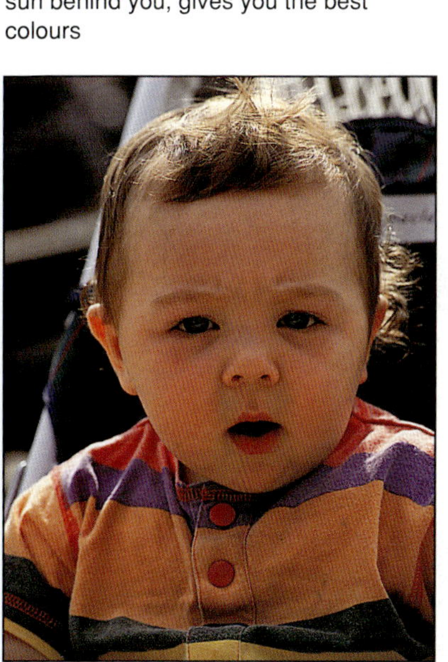

▲ **Figure 12** Backlighting: no sunlight falls directly on the baby's face; instead it is lit by indirect, reflected light which lowers contrast and colour saturation

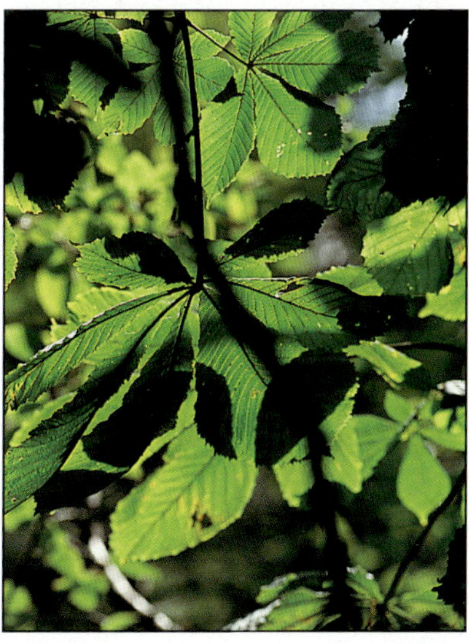

▲ **Figure 13** Backlighting is ideally suited for translucent subjects, such as leaves, as it shows their colour and form at their best

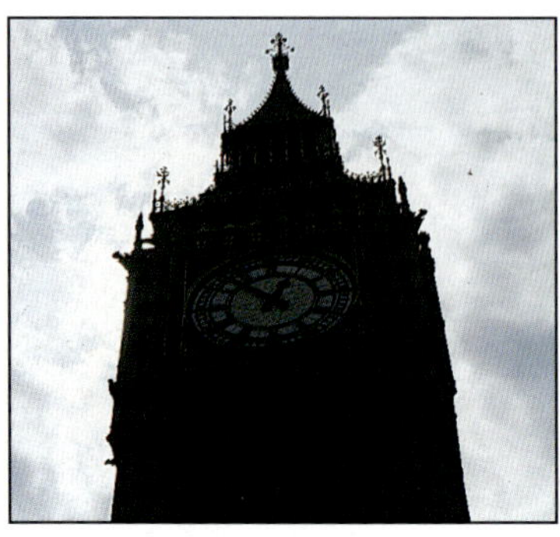

Figures 14 and 15 Shooting into the sun can mean that your subject turns into a silhouette (left). Using the backlight button (below) improves the exposure on the subject, but the background becomes overexposed and the colours look drained

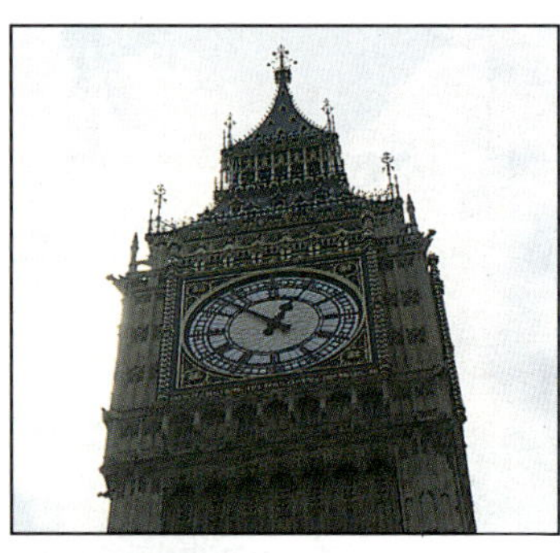

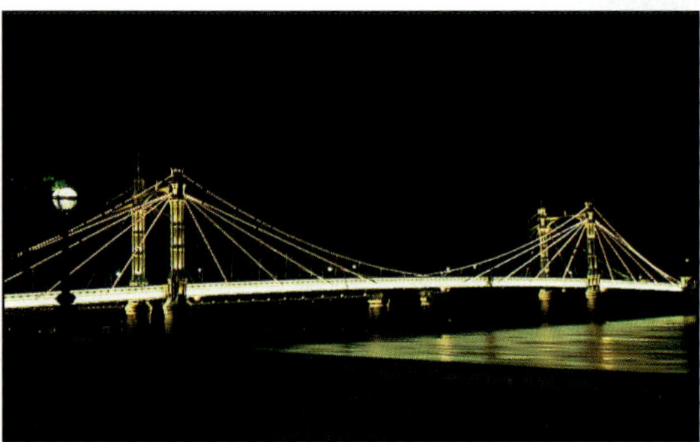

Figures 16, 17 and 18 A manual iris facility gives you complete control over exposure. This is particularly useful for high-contrast scenes, such as cityscapes at night, where correct exposure is a matter of taste. Do you expose for the shadows, or for the bright lights? Or go for something in between? Which of these three exposures do you prefer?

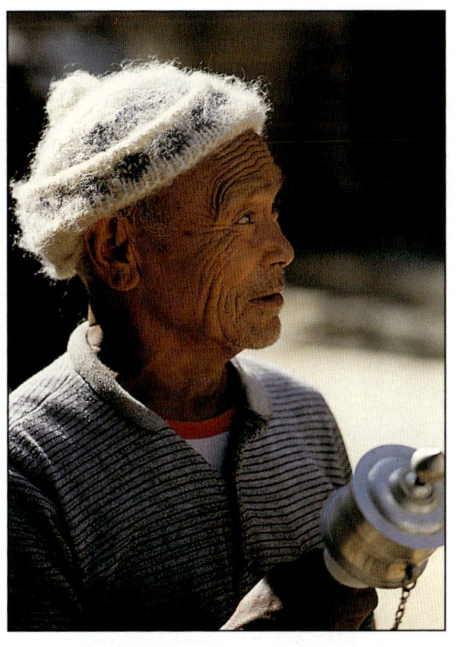

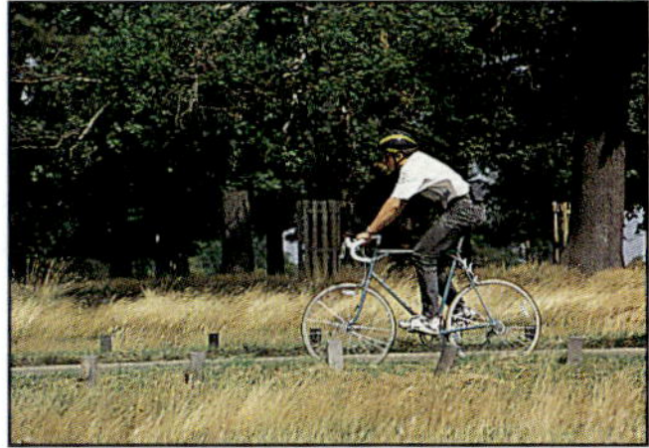

◀ **Figure 19** Looking space: when shooting a person who is looking offscreen, always frame them so that there is more room in front of them than behind, otherwise the shot looks cramped

▲ **Figure 20** With subjects moving across the scene leave more room in the frame in front of them than behind ...

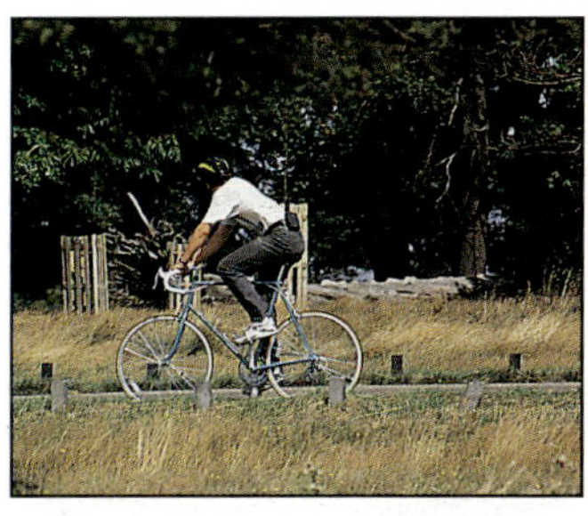

▲ **Figure 21** ... doing the reverse, by leaving more room behind than in front, makes the shot look unbalanced

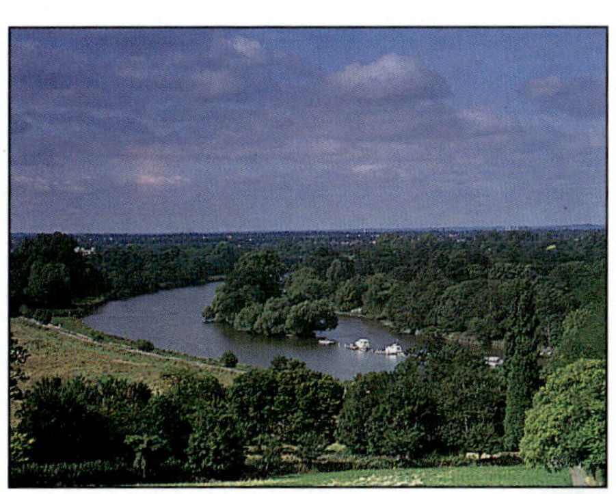

▲ **Figure 22** Natural curved lines, such as those on a river, lead the eye gently through the shot

▶ **Figure 23** Diagonal lines in your composition make the shot more dynamic – forcing your eye across the picture

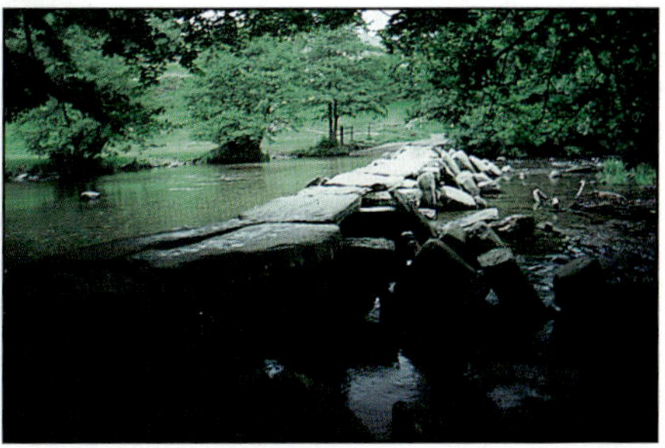

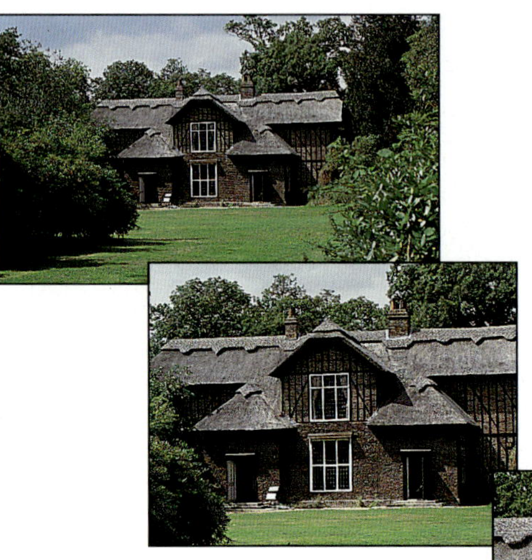

Figures 24, 25 and 26
Zooming in: by starting in the same position and using your zoom to move you in 'closer' to the subject, you enlarge the background with the subject, forcing it forward in the picture

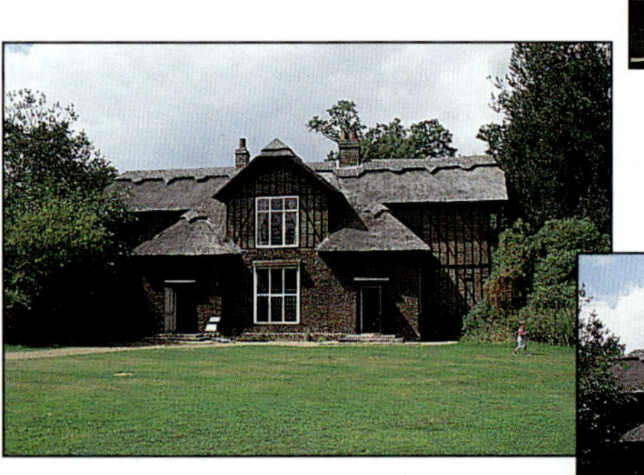

Figures 27, 28 and 29 Tracking in: the alternative to using your zoom is to move in closer to the subject. This looks more impressive as the perspective changes as you move in – forcing the foreground to become more dominant, and the background to recede, or even disappear, from the shot

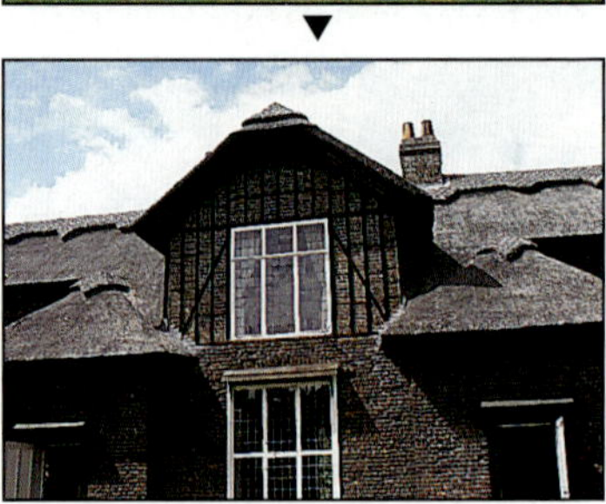

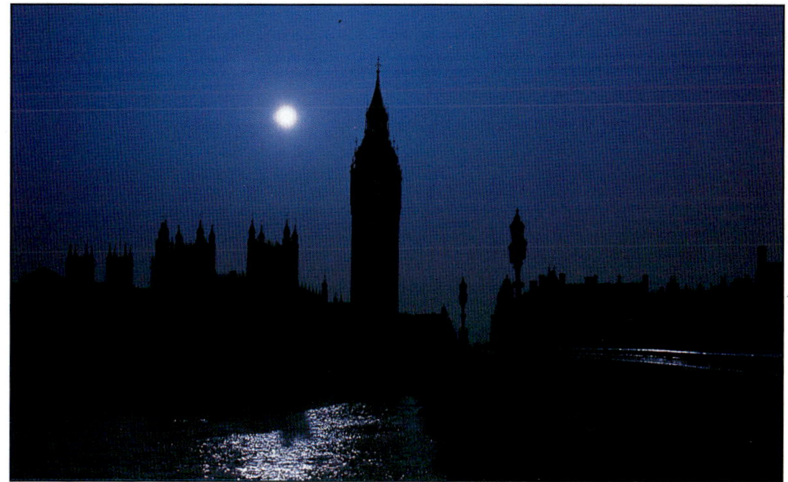

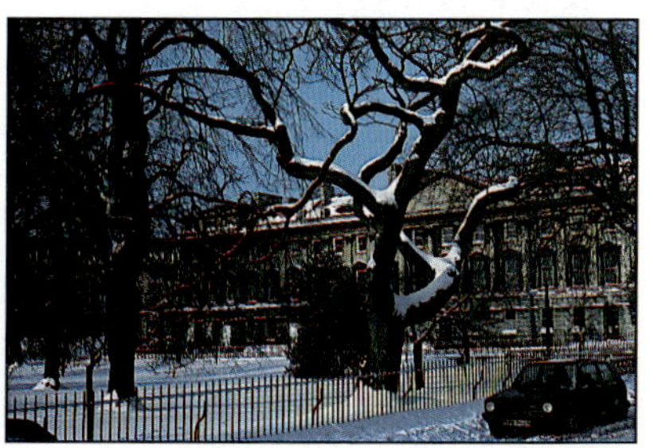

Figures 31, 32 and 33
Digital mix: this useful camcorder feature allows you to fade one shot directly into the next. As one shot gradually disappears, the next gradually appears. The camcorder does this by memorising a still of the previous shot, and superimposing the new, live shot on top of it

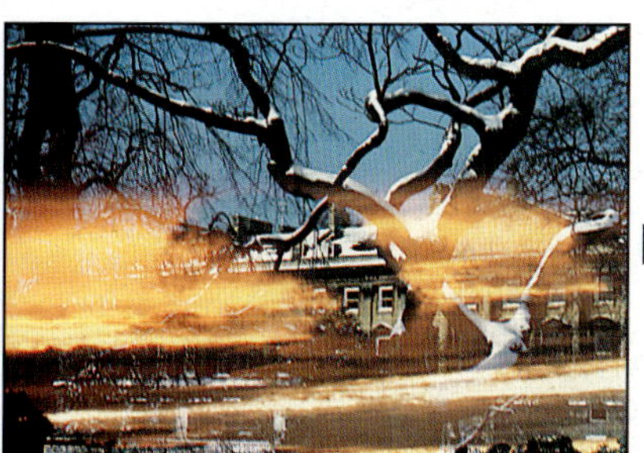

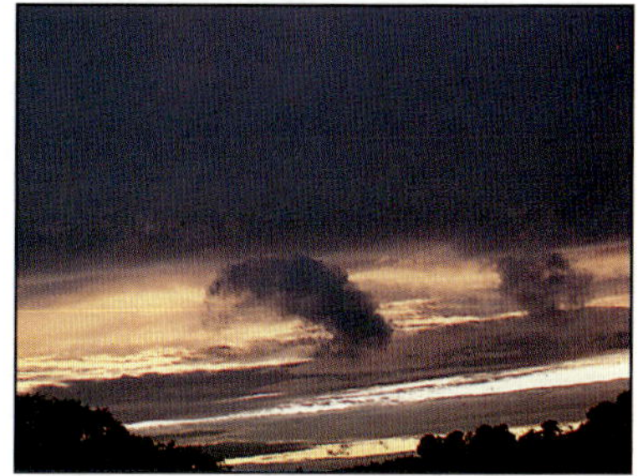

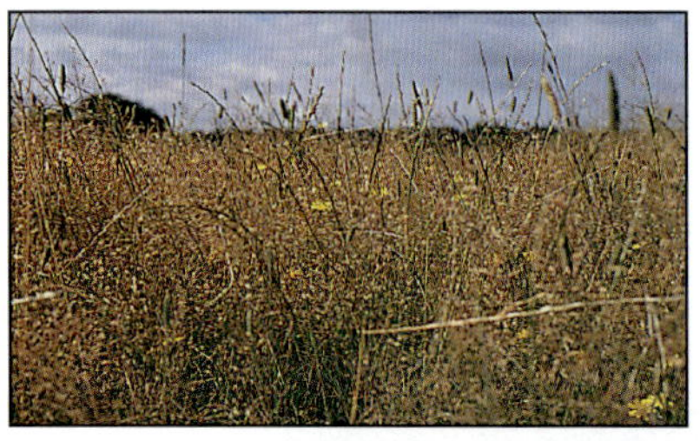

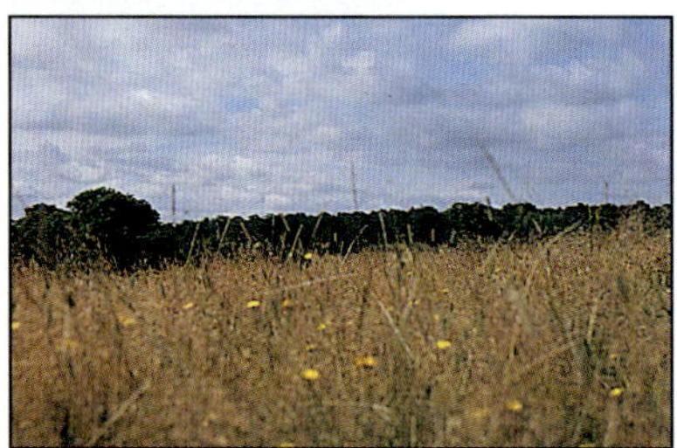

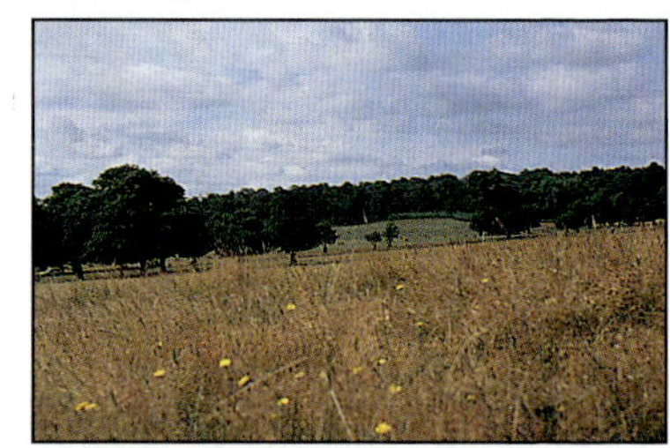

▼ **Figure 37** Dutch tilt:
turning your camcorder
slightly at an angle gives an
interesting effect that is
useful for shots of cars and
skyscrapers

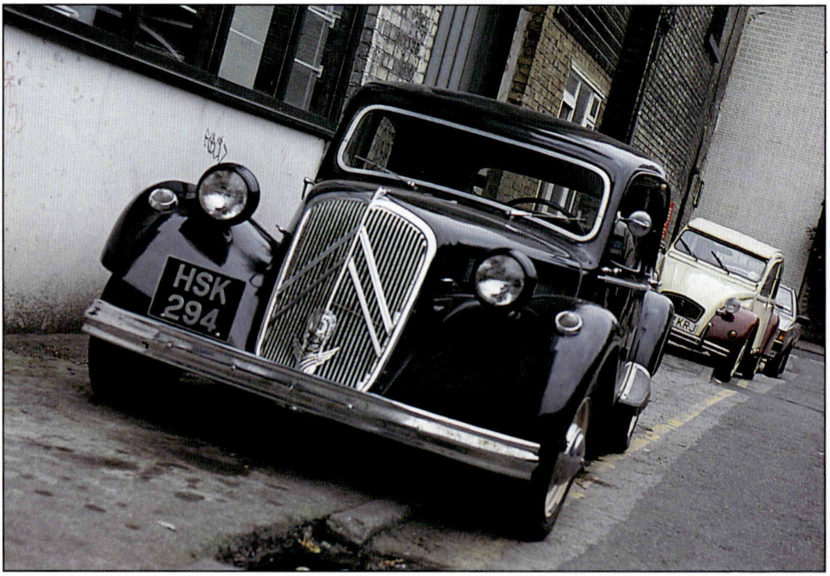

times when, for variety or necessity, you will need to point the lens up or down or to lower or raise the camera significantly.

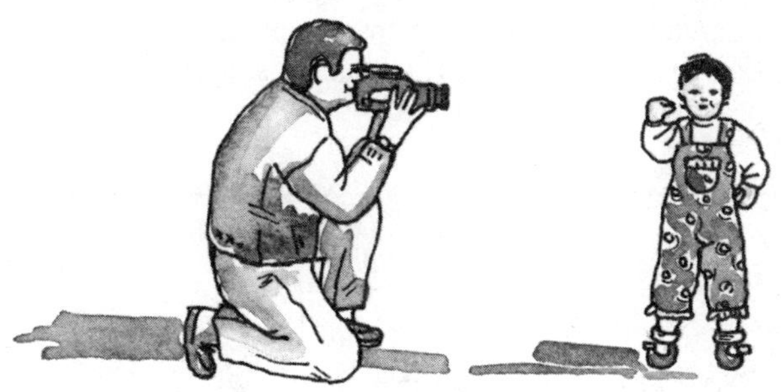

Kneel to get to the eyelevel of a toddler

When shooting people, you will normally want to keep the camera at their eyelevel. If they are the same height as you this will be easy, as the camera is normally held at your eyelevel. When shooting children, however, it is a natural tendency, just to point the camera downwards. This has the unfortunate effect of affecting the perspective, so their heads look disproportionately bigger than their feet. To the viewer it looks, not surprisingly perhaps, that you are looking down on them – exaggerating their small size. On occasions you may want to do this – but for normal-looking shots you should get down to their level to shoot. In the case of a five-year-old, this might mean kneeling. For a baby, it may mean lying on the ground yourself. The same technique should also be used when videoing pets, such as cats and dogs.

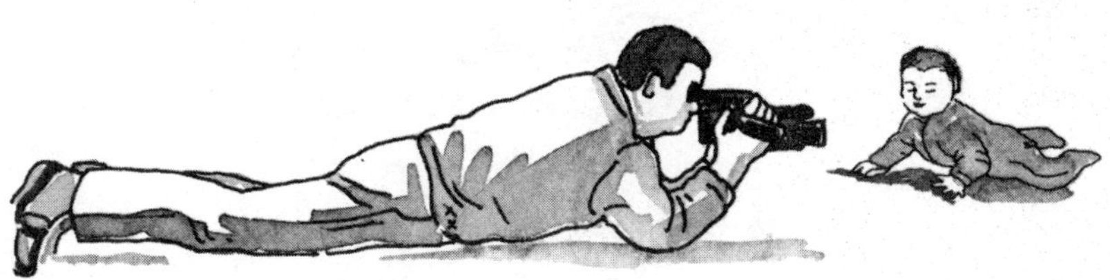

Lie down to get to the eyelevel of a crawling baby

Just as you can make people look smaller and more demeaning by shooting them from a higher position, you elevate their status by using a low-camera viewpoint. If you kneel down and shoot up at someone, they automatically look more powerful and authoritative. It is a good trick to use, again, when shooting children. Kids love to see themselves portrayed in this way – making them look king of the castle. The same technique is often used when shooting tower blocks – as you often have little choice but to point the camera upwards. The converging parallel lines that result accentuate the building's dominance – although, in this case, at least this is what we are used to seeing.

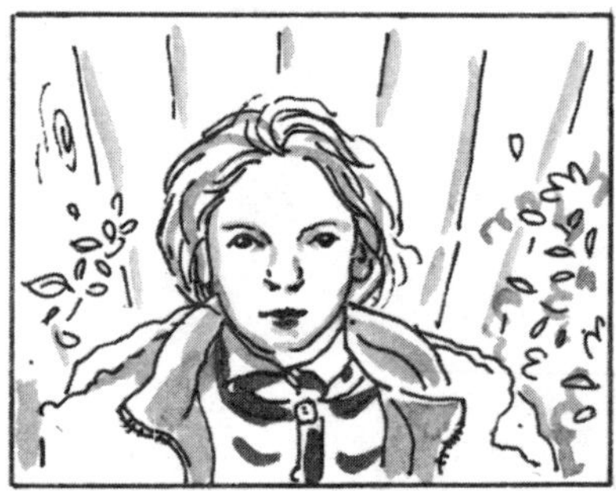

High angle – shooting down on subjects, emphasises their small size or demeans them

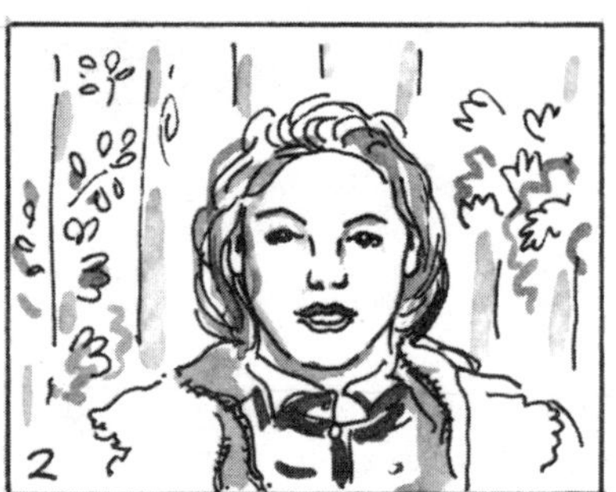

Normal angle – camera level matches eyelevel of subject

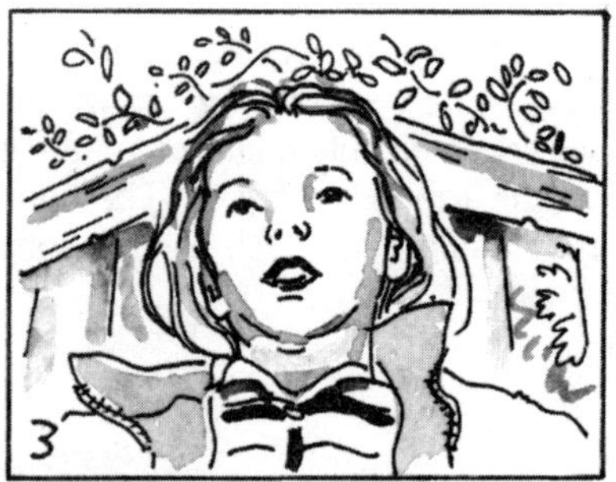

Low angle – camera looks up at subject, this exaggerates their height or importance

10
THE MOVING CAMERA

One of the great advantages of the camcorder is that it can be moved while it is recording. In this way, it can closely imitate the way we normally see things by moving our head, or taking things in as we walk. Camera movements can help bring a video alive – often adding that third dimension that is often missing from our two-dimensional TV sets.

There are two broad types of camera movement. There are those from a fixed point – where all the camera does is swivel from side to side, or up and down (known as *pans* and *tilts*). The other type is more adventurous, and involves the camera completely changing its position as it records – either by moving the camera towards or away from the subject, taking the camera around the subject, or changing the height of the camera. These are known as *tracks*, *crabs* and *cranes*.

All of these camera movements have their particular uses, and are executed in slightly different ways.

Panning

A pan (or panoramic shot) involves swinging the camcorder through a horizontal arc – from right to left, or vice versa. The alternative to the pan is to use a wider shot from further away, or using a wider lens setting – but this usually means adding unwanted foreground detail at the expense of detail in the subject itself. A panning shot has three distinct uses in movie making:

- To help show objects that are too wide to fit into the frame sensibly – such as a hilltop vista, or a big building.

- A pan can also be used to show the relationship between two different subjects. A good example of this is with an action – reaction shot. Grandmother enters the room carrying a big gift-wrapped parcel (the action) – you pan round to a shot of the child's beaming face, as he realises that the present is for him (the reaction).
- The most obvious use of a pan is to follow movement – whether it be a person walking across the lawn, or a racehorse galloping across the frame.

Pans are notorious for being far more difficult to execute than it might seem. It is essential that they are performed as smoothly as possible. They must also be slow enough to give time for the viewer to see what you are panning across.

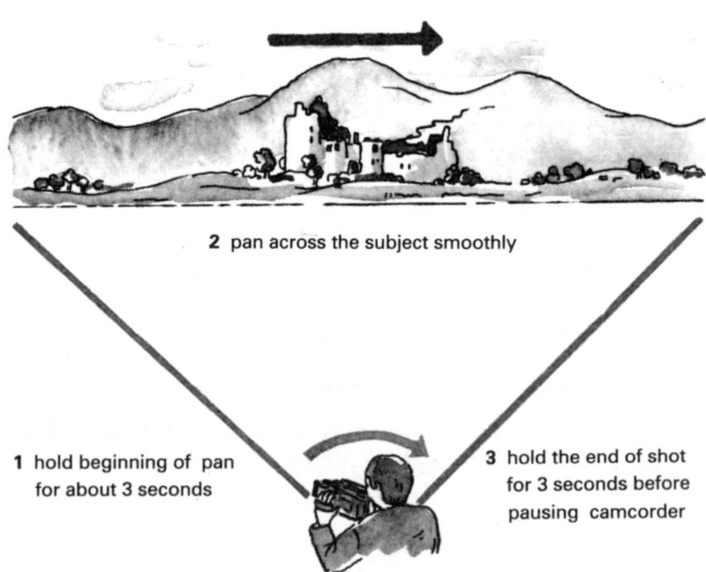

How to pan

The smoothest pans are done with a camcorder fixed to a tripod with a fluid pan-and-tilt head. However, you will undoubtedly have to do without this luxury on occasions. The secret of a smooth handheld pan is to practise the movement before you commit it to video. It is all too easy to find that you can't quite reach where you want the recording to stop without staggering over sideways.

The correct technique to use is to hold the camcorder tightly in your hands with your feet directly below your shoulders. You then twist your body from the waist upwards to the extremity of your pan. You then untwist your upper body gradually – then twist it in the opposite direction. By practising this before you record it, you can ensure that you are not going to overstretch yourself to get everything in the shot – and that you are not going to overshoot the finishing point. In practice, you will find that you can only safely pan through about 90° in this way. Never pan back directly after panning forwards – this is the height of amateurism!

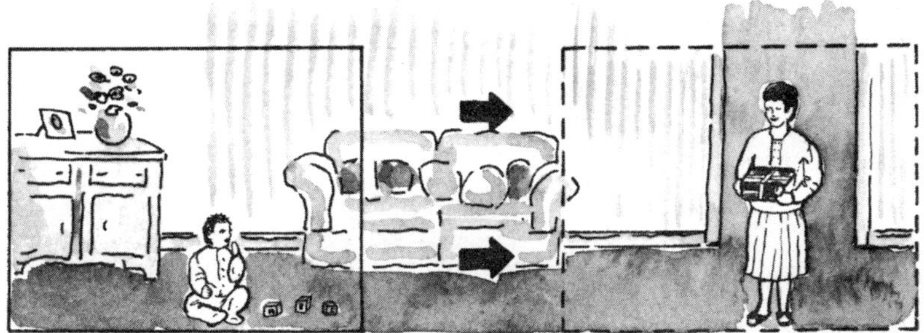

Panning to show the spatial relationship between two people in different previous shots

You should try to start and finish a pan with a stationary shot lasting from three to five seconds. If you are shooting a large building, for instance, start with a three-second shot of one end, pan across, and finish with another three-second shot of the other. The reason this is necessary is that it is very disconcerting to cut from a moving camera shot to the next shot. All camera movement shots should ideally always have a stationary introduction and conclusion.

When panning to follow a moving subject, you should remember to leave more space in front of the subject than behind (moving space, see previous chapter). Starting and finishing with a stationary image is more difficult – but is still possible. Your first shot is likely to show the subject moving towards the camera, and therefore requires no camera movement until the subject gets closer to you. You then pan, and to finish the shot, you can stop the camera, and allow the subject to exit the shot at their own speed. This should allow you to get your three-second beginning and end to your pan.

Tilting

In essence, a tilt is the same as a pan, except that it is a movement in the vertical direction. Its main use is for fitting tall subjects into a single shot without a lot of wasted space. For obvious reasons, it is less useful for moving from one subject to another, or for following movement.

A tilt is carried out in much the same way as a pan. The important thing to do is to rehearse the movement before recording – to make sure that you do not outstretch yourself. Again you should aim to start and finish the movement with a stationary shot lasting three or more seconds.

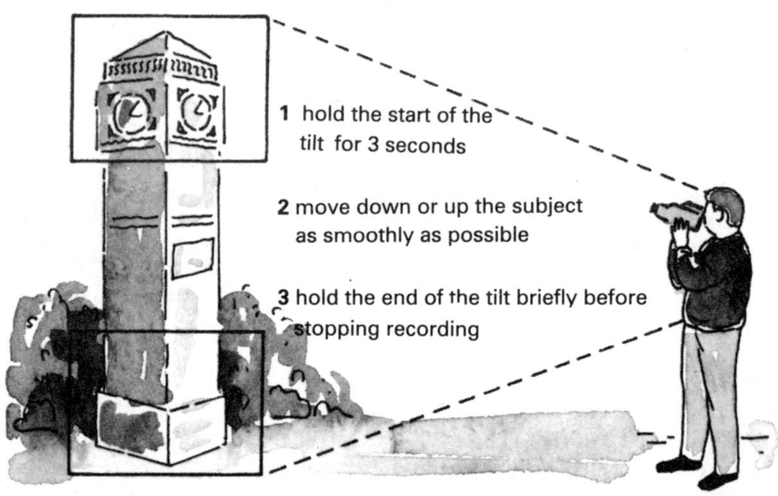

How to tilt. Practise the movement before you start – ensuring not to stretch through more than a 90 degree arc

One problem that is peculiar to tilts is that of exposure. When swinging up from the ground to the top of a bell-tower, for instance, there is almost bound to be a significant brightness difference between the start and finish. There are two ways to cope with this. The first is to leave the camera on auto-exposure and hope that the sky is not so bright that it turns the bell-tower into a silhouette at the top. The second is to use manual iris control to ensure that the tower itself is properly exposed and then put up with the fact that the sky is likely to look washed out as you tilt up.

Tracking

A tracking shot is one in which the camera moves forward, or back, while recording. In some ways, it performs the same job as a zoom – changing the image size of your subject. However, it has great advantages over a zoom shot. When you zoom in on a subject, the perspective does not change – and because of this the background looks squashed up and unnatural (see Chapter Seven). When, instead, you track forward towards your subject, the subject still becomes bigger in the viewfinder, but the perspective in the picture is constantly changing. Instead of the background seeming to move towards the subject as you zoom, as you track the background seems to stay at the same distance. With a zoom, the background seems to get bigger and you end up with only part of the original background filling the frame. With a track forward the background remains the same size throughout.

A track forward, therefore, has the advantage that you are not accentuating the background. It is thus seen as being preferable to the zoom in professional circles, as it looks far more dynamic.

An alternative use for a tracking shot is to follow a subject as it moves across your field of vision. In this way the camera mimics the subject's movement exactly – and the background to the shot is constantly changing.

A tracking shot gets its name because in the world of professional film-making, it is performed by laying down a miniature railway track – which the camera moves along on wheels as its films. This ensures that the movement is extremely smooth. Unfortunately, such preparation and expense is out of the question for most camcorder owners. The problem, therefore, is how to make tracking shots which are not spoilt by the camera bobbing up and down. Walking with the camera as you are recording rarely looks impressive (for an exception to this see the section on crabbing below).

There are several ways in which you can get a relatively smooth tracking shot without spending a fortune:

- A dolly – a set of three wheels fixed to a Y-shaped frame, that can be attached to the bottom of most tripods. This works well if the ground is even, with no bumps.
- A wheelchair, supermarket trolley, or pram – all of these can form an improvised dolly cart that you can sit in as you are filming. You will need an assistant to push you. How smooth the ride is will

depend largely on the size of the wheels – the bigger the better. Also make sure that the wheels are well-oiled – otherwise they can make a hideous screeching noise that will spoil the soundtrack.

- A Steadicam J.R. – an ingenious American invention that when attached to your camcorder counterbalances it so well that the whole unit can be supported with one hand. The unit dampens any movements, so that they look very smooth on screen. The user watches a small TV screen on the Steadicam, so that you don't touch the camcorder directly at all. It is not cheap, costing about £600 (or $600 for the American version).

Whichever system you use, you will find that the only way to make tracking shots look effective is to use the wideangle end of your zoom. This not only helps minimise the effect of camera shake as you move, but also gives you some margin for error to ensure that the subject remains in the frame.

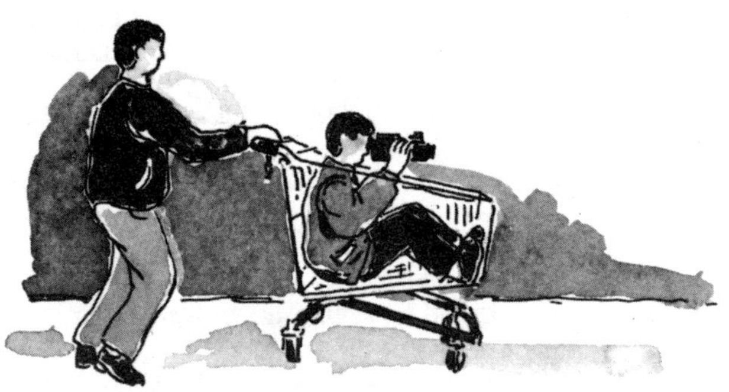

A supermarket trolley can be used to provide a smooth tracking shot

Crabbing

Crabbing is a form of track shot that allows you to walk with the camera for short distances. It is primarily used for tracking round a subject – such as a statue. The basic principle is that you walk sideways, taking very small steps at a time – avoiding the camera bobbing up and down.

First you stand with your knees slightly bent, and your feet about a foot apart. To move you make a two-step movement. The first step is to swing one leg in front of the other so that you are now standing with your legs crossed. The second step is to bring the leg behind you to the side – so you are again standing with your legs uncrossed, with your feet about a foot apart. You then continue this two-step movement as far as is necessary.

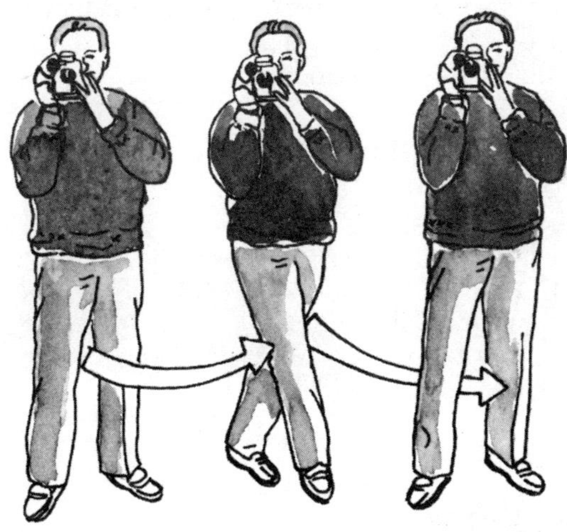

How to crab

The technique can take a bit of practice to make sure that the camcorder is moving smoothly throughout. Do not move too quickly or try to make large steps as you will lose your balance slightly, and your video will look shaky. Although there is no need to rehearse the movement beforehand – you should check in advance that there are no obstacles in your intended path!

Craning

A crane shot is a track shot in the vertical direction. It gets its name because, to achieve this shot in the movies, a hydraulic crane is used to lift the cameraman into the sky. For the amateur, crane shots are usually restricted to raising the camera level by just a couple of feet – but this can still be effective if used in the right situation.

A simple form of crane shot might be to identify a person gradually in a shot. You might start with a big close-up of a person fiddling with a pencil at a desk – and then gradually crane up to reveal the face of the person (showing their identity and expression).

A more elaborate form of crane is to engineer the shot so that it starts with only foreground, and, as the camera rises, a wider view shows you exactly where you are. For example, as an introduction to a video about your holiday in Athens, you could start with a wide-angle close-up of a flower. As the camera rises it looks over the flower to show the Acropolis in the distance.

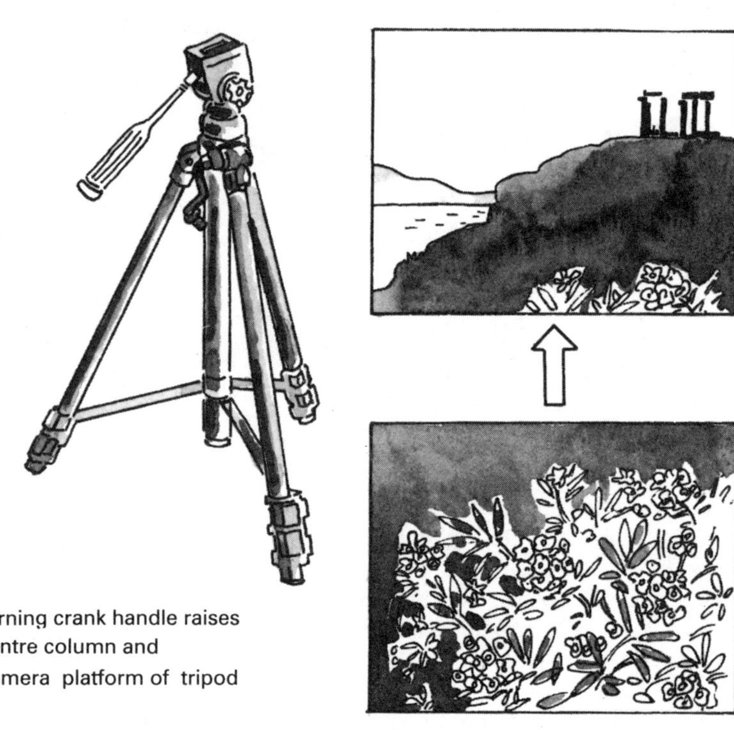

Craning up: shot starts low with flowers filling the frame. As the tripod platform rises the shot looks over the flowers to the landscape in the distance

A reasonably effective way of executing a crane shot is to squat down and then simply to stand up as you begin recording. Again it is a good idea to start and end the sequence with a static shot lasting three or more seconds. For smoother-looking results use a tripod that has a geared centre column. This allows you to extend the height of the camera platform without moving the legs of the tripod, by means of a crank handle.

11
— CONNECTING THE — SHOTS TOGETHER

Although a video is made up of a number of individual shots, it should not end up like a photo album, where all the shots can be completely unrelated. A video tells a story – and because of this the viewer expects there to be some connection between one shot and the next. Normally the connection is obvious, but sometimes it needs explaining. You do not want the viewers to ask themselves 'What on earth has that got to do with anything?' (unless you are building up suspense and are deliberately teasing them). There must therefore be some logic in the way that you put your sequences of shots together.

Thanks to the huge exposure that we have all had to film and television, most of us have grown used to the 'language' of movie-making. But although we understand it – most people cannot use it themselves. No one has explained the rules and conventions to us – we just accept them on screen – and are confused when they are not obeyed.

For instance, we willingly accept that a TV documentary can cover the events of several days – or even years – in just 30 minutes of video. We readily accept if one shot shows the outside of a house and the next shows someone sitting in front of a living-room fire that that person is inside the house we saw in the first shot. Similarly, if we see a car driving across the screen from left to right in one shot, and in a following shot we see the same car travelling from right to left, we assume that the car has gone somewhere, and is now coming back. These are just some of the conventions of movie-making that we have learnt – although we may not know the rules that the camera operator has used in order to get us to accept these jumps in logic.

In the rest of this chapter we will look at some of these rules and how they can help us tell a story on screen coherently and efficiently without having to shoot absolutely everything that we see.

———— Condensing time ————

The commonest criticisms of amateur videos is that they go on far too long. Each shot goes on for several minutes, with no apparent thought as to whether what we are being shown is interesting or not. The camcorder user must be selective in what is shown on screen – and then must cut it down to the shortest length possible, without the viewer feeling that he has been rushed.

The key to making your videos short and succinct is to break down what is going on into shots – as we have seen in the last two chapters. But this technique also means that you have the chance to change the picture – by altering the shot size or camera angle. This variety helps keep the viewer interested in what is going on – as there is always something new to look at, even if the subject is the same.

The important rule to remember about shot size when putting sequences together is that you should always include a wide shot (normally a long shot or extreme long shot) whenever you change locations. This tells the viewer that you have moved between shots – even if it is just another room in the same house. It does not necessarily have to be the first shot in the new location, but it must be used early on. Once the new location has been established, you can concentrate on a succession of mid-shots and close-ups.

Sometimes it is difficult to see exactly where you can cut a scene without losing something vital. The kids are playing football on the lawn and each moment looks just as good as the last – so why stop recording? The answer is that your video will look better with ten different shots of ten seconds than one shot lasting 15 minutes – so force yourself to use the stop/start method, even if it seems inappropriate.

Conversations are even harder to cut – they flow from one person to the next, building on what has gone before. The answer here is to try to catch individual sentences completely – then break off. Topics of conversation do not usually change that quickly – so your next recorded

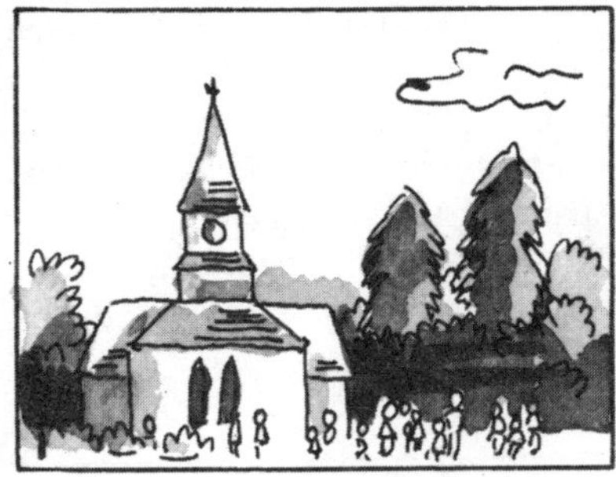

One of the first shots in a sequence should establish the location – here an extreme long shot of the church and wedding party

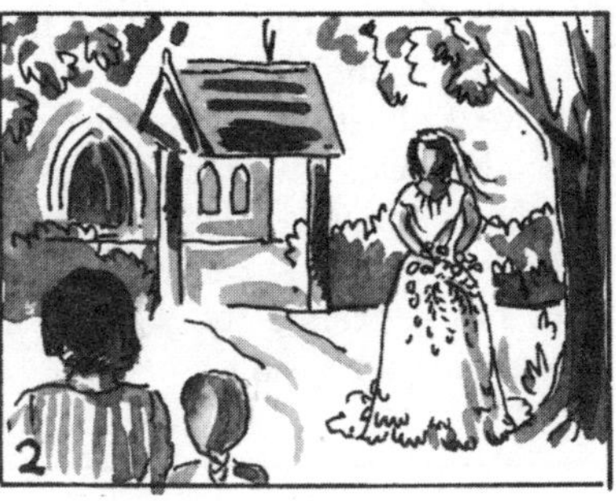

A different angle and a closer shot – here a long shot of the bride

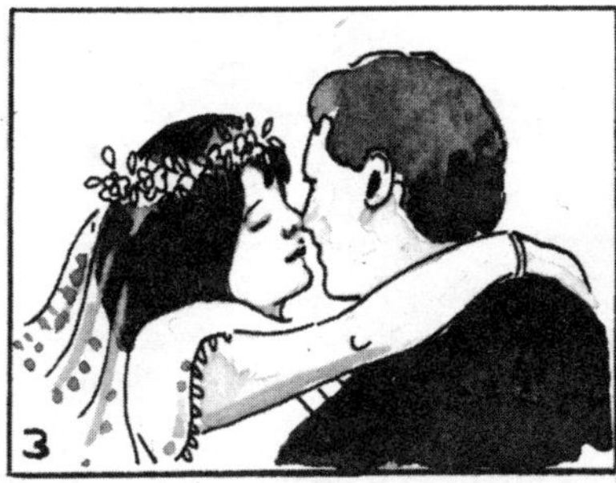

Another change of size and angle – here a close-up of the couple kissing

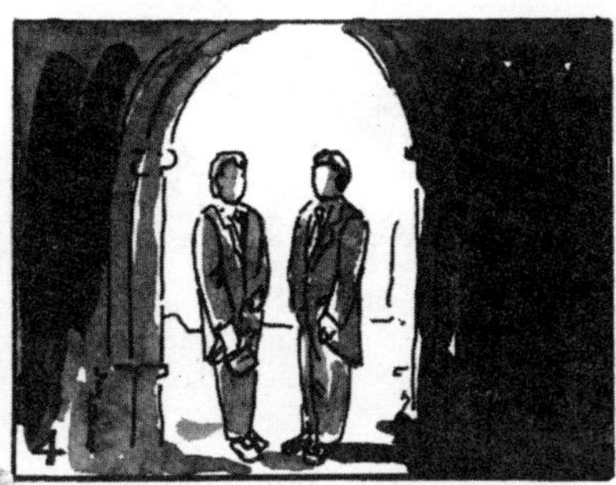

Another change of position and shot size – here a long shot of the ushers from inside the church looking out

A typical sequence of shots showing how camera angle and shot size change from one shot to the next

sentence probably will not sound out of place if you pick your moments well.

Jump cuts

One of the main reasons for changing shot size and camera angle between shots is to avoid a jump cut. A jump cut is where two successive shots are so similar that the subject seems to 'jump' on screen. It looks like a mistake – as if a part of the video has inexplicably disappeared. For example, you are videoing a child sitting at a desk painting. You stop recording, and a minute or two later start recording again from the same position, with the same zoom setting. Much of the picture is still in the same place in the frame – the desk, the background, etc. The child, however, will have inevitably moved slightly – the hands will be in a different place, and the painting will be that much nearer completion. The small changes are highlighted because so much in the shot has remained the same – and the viewer will be reminded that something is missing.

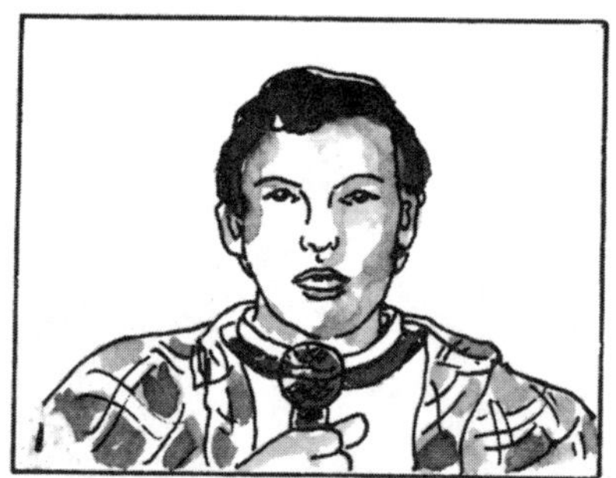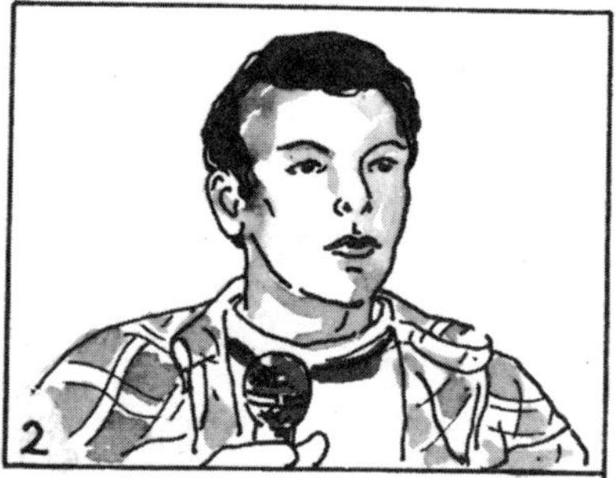

The jump cut. Cutting between shots 1 and 2 is a mistake. The shot size and camera angle are the same. The small change in subject position is therefore made all the more obvious on screen as the image appears to 'jump' from one position to the other

Ideally, to avoid any chance of a visible jump, you should change both shot size and camera angle between shots. Changing just one of them slightly still runs the risk of looking wrong on screen – and making significant change to just one of them can look a bit forced.

On some occasions you may not be able to change your shooting position and making straight cuts will inevitably involve a jump in the action. The solution here is a *cutaway* – a shot of something that is not in the main shot, but is somehow connected with it. With the above example, a suitable cutaway might be a shot of previous paintings hanging on the wall – or a close-up of the mother looking on with pride. This provides a visual break that then allows you to return to your main subject without an on-screen jump.

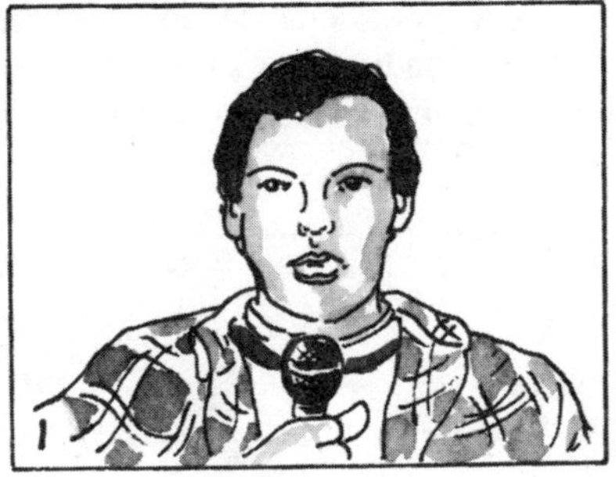
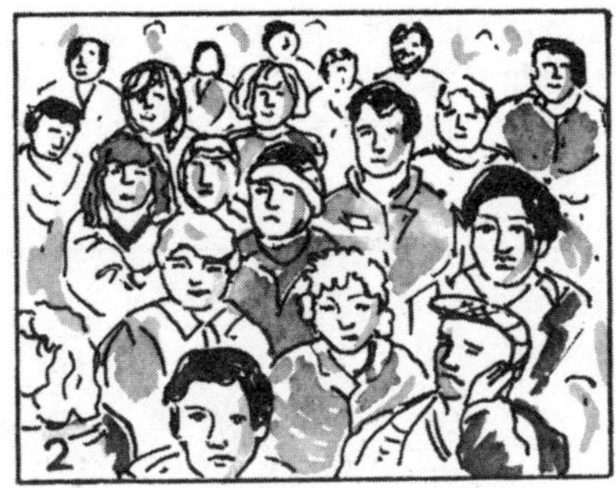
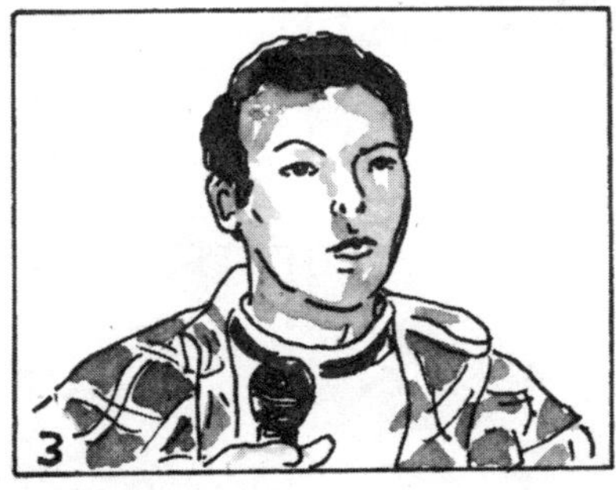

To avoid a jump cut insert a 'cutaway' between the two similar-looking shots. A cutaway can be of anything that is relevant to what is going on in the other shots but is not contained within them. Here a shot of the speaker's audience suffices to make a break between shots 1 and 3

Continuity of direction

One of the fundamental rules of video is that when you are shooting a moving subject they should appear to move in the same direction in each shot. For instance, if you are videoing the family on a Sunday afternoon walk, they should always appear to move from left to right across the screen (or always from right to left). If the direction changes, the viewer's natural assumption is that they are now heading back from where they came.

If you are not careful, it is easy to end up with a video where the direction of movement is continually changing. The secret is to shoot from the same side of your subject throughout – and avoid the temptation to swap sides to get a more interesting view. Some textbooks describe this rule by saying that there is an imaginary line that your subject is heading along – and at all costs you should avoid crossing this line. It is also known as the 180° rule – you can move from in front of your subject to behind them, but you cannot cross their path.

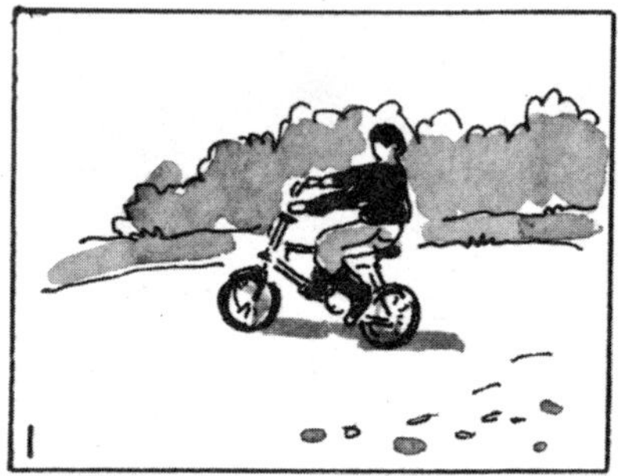
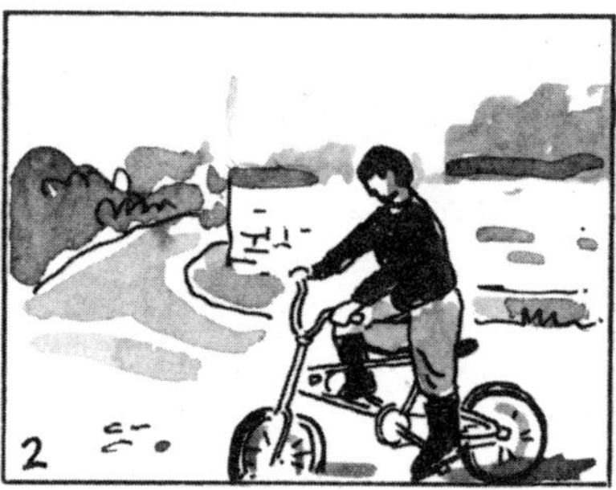
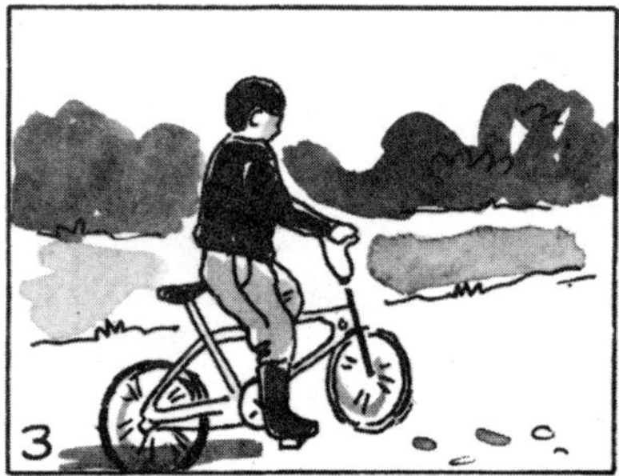
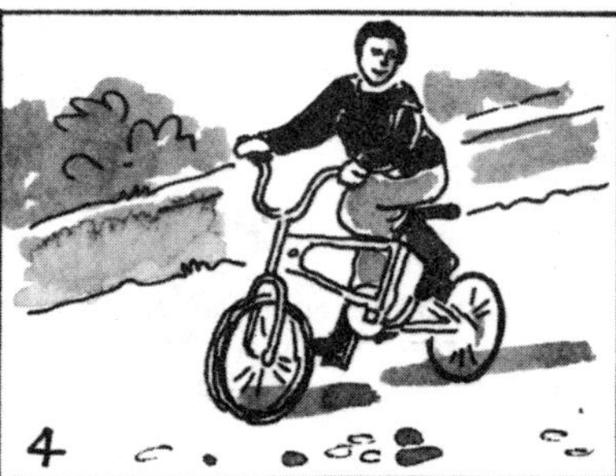

Continuity of direction. Spot the odd shot out! The bike seems to have changed direction in shot 3 for no apparent reason. 'Crossing the line' to look at subjects from the other side can be extremely confusing on screen.

Sometimes you may have no option but to 'cross the line' and to change the direction of on-screen movement. If you really need to do this, there are two possible solutions that will make the change more acceptable to the viewer (although neither is perfect):

- Keep recording as you cross the line of action – so the walkers, say, look as if they are heading towards the camera, then move to the other side of the path as they approach.
- Insert a cutaway between the two opposite shots. For instance, at a football match, a brief shot of the crowd, or a shot of the scoreboard, could be used to provide visual relief from the main subject – and give you a chance to change on-screen direction.

Continuity of direction does not only affect action footage – it also has its role to play in more static subjects. Most importantly, if a person is looking from left to right in one shot, they should be looking in the

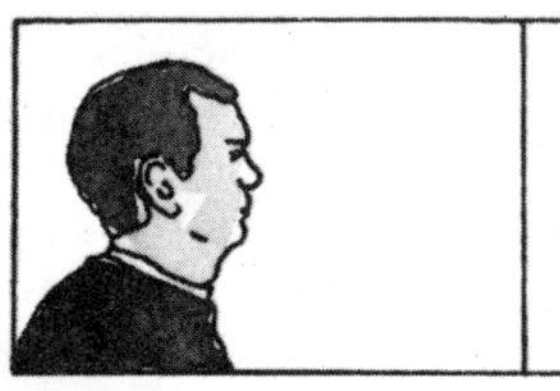
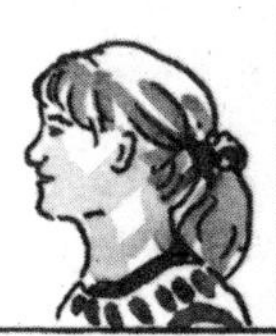
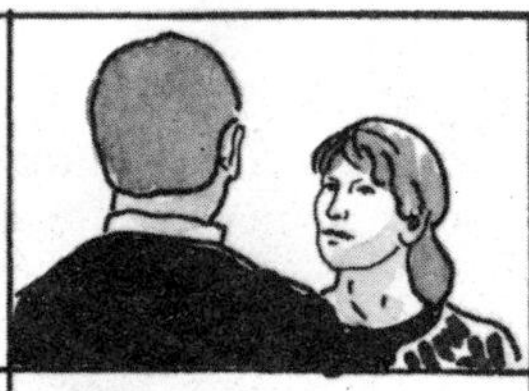

In these shots the man stays on the left of the screen looking right and the woman on the right looking left

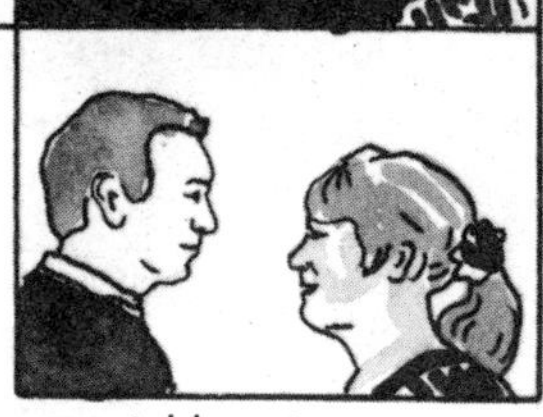

acceptable cuts

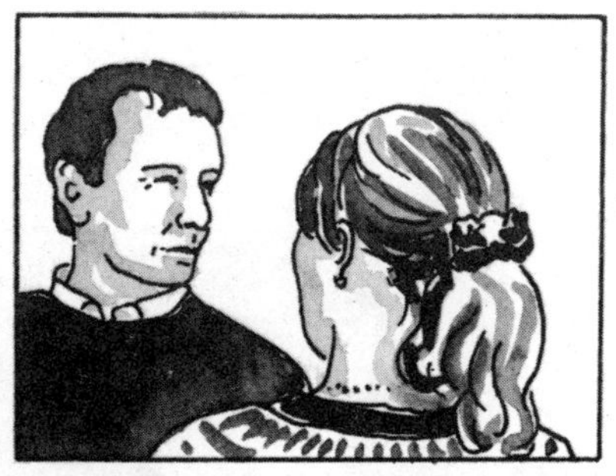

original shot

In these shots the camera has crossed the line and the man and woman appear on the wrong side of the screen and are looking in the wrong direction to match with the original shot

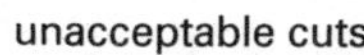

unacceptable cuts

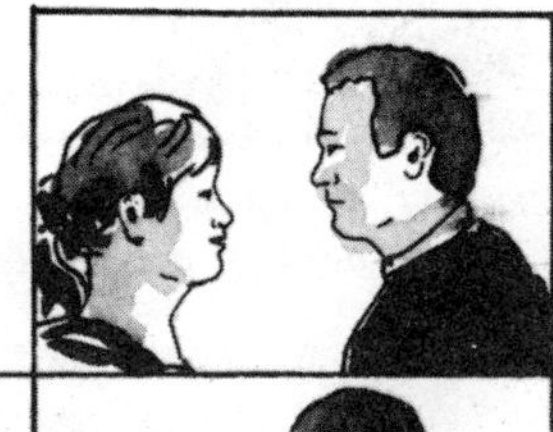

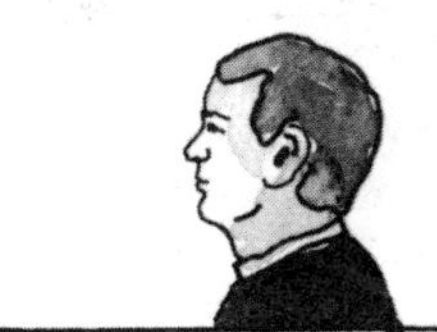
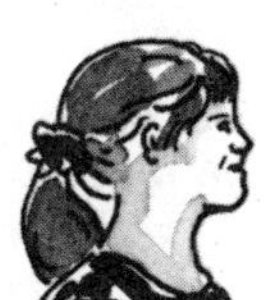
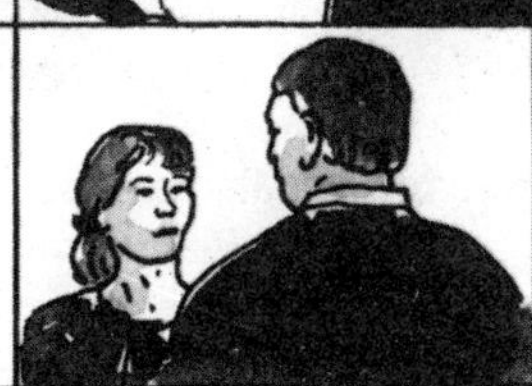

same direction in the next. You can imagine the confusion if you did not obey this rule when you were shooting two people talking to each other. A is looking screen right at B, and B is looking screen left at A. If the camera goes round them to the other side of them, A is now looking to the left, and B is looking right. They could seem to change the way that they looked every time they opened their mouths! Again, the secret is to think of an imaginary line on which both people are standing and which the cameraman must not cross. This convention is known as matching eyelines. By having two successive shots of people looking in opposite directions, you are telling the viewer that the two people are interacting with each other in some way. For the effect to work best, the shot size used for each should be roughly the same – and the amount of headroom (space between the head and the top of the frame) should be the same.

Using shots in pairs in this way is a common way of connecting shots together. The second shot is known as the reverse angle shot of the first – and together they form a complementary pair. But the same technique can be used even when you are shooting just one person. If the person is looking off-screen, we assume that the next thing that we are shown on screen is what that person is looking at. Reverse

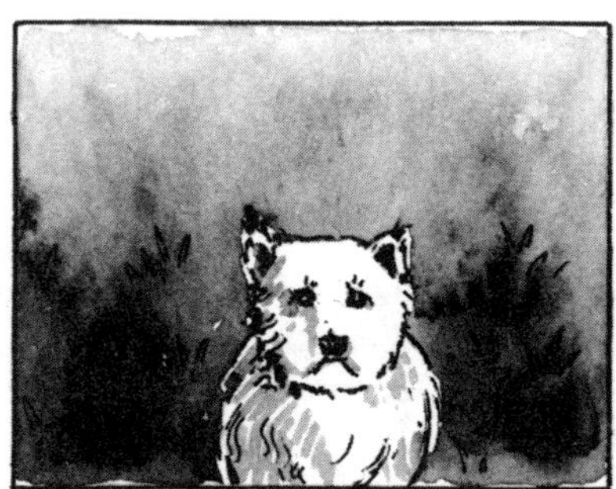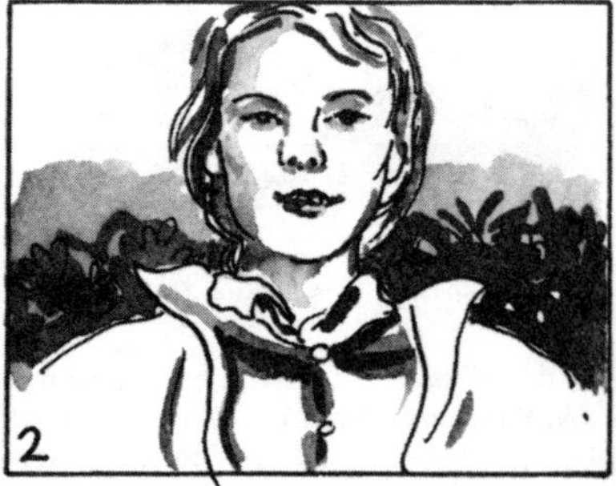

A reverse angle shot is the natural partner to shots of people – or animals. Here a shot of a dog looking up is followed by its owner looking down at it. Using a high camera position for shot 1 and a low angle view for shot 2 helps to reinforce the difference in size between the two subjects

angles can also be used to reinforce height differences between subjects – a shot looking down on a cat can be matched with a reverse angle shot of what the cat is looking up at, with the camera shooting up from the ground.

A variation on the reverse angle shot is the point-of-view shot (POV). Here the second shot is taken from precisely the same place and in the same direction in which the person in the first shot is looking.

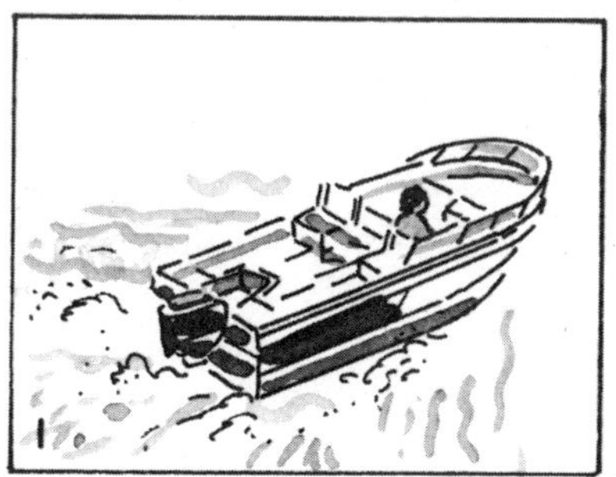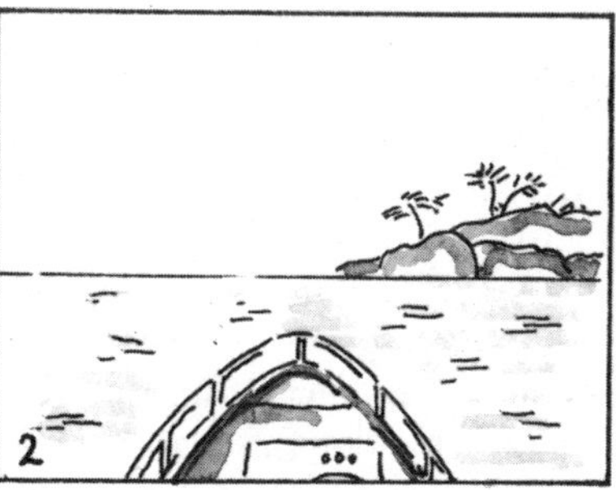

A point of view shot shows you precisely what the person in the previous shot can see. Here we cut from a man driving a speed boat to a shot of what he can see

Cutaways

As we have seen, cutaways can form an important part of a video. They are a visual break helping you to avoid jump cuts and confusing changes in screen direction. Remembering to shoot cutaways in the right place can be difficult – especially if you do not want to miss any of the action. If you are going to edit your video at a later date, you can shoot your cutaways as you see them, and when you have the opportunity – so they can be inserted in the right place at the editing stage. For instance, if you are shooting a wedding, suitable cutaways that could be shot at any time during the day might be: details of the church, close-ups of floral displays, shots of the cake, and close-ups of the order of service. With practice, you should be able to make sure that these are shot in roughly the right place to be of most use in the final video.

Remember that cutaways do not have to be of inanimate objects. Shots of on-lookers can provide the most useful visual break. These are often called reaction shots – showing people's reactions to what is going on in the main sequence. So cheering spectators at a sporting event, of smiling guests at a wedding, and of pleased parents at a school concert are essential shots.

Parallel action

Normally a video will concentrate on one sequence of events before moving on to another. However, there are times when you might want to intercut between two separate sequences to show that they are occurring simultaneously. It is a technique that is commonly used in thrillers. We see the heroine peacefully at home preparing a meal. Periodically this sequence is interrupted with another sequence of the evil villain creeping round the garden carrying a knife. We know that the two lines of action will meet at some point and a feeling of suspense is created as we wait for the inevitable.

This technique is called parallel action, but it is not only useful for creating tension. It can simply be used to show two different stories at the same time. An everyday example might be to show two different activities going on in the same place at the same time – children's games in the garden intercut with the adults talking in the front room. By recording, or editing, the sequences together in this way you reinforce the fact that both events are taking place simultaneously and help ensure visual variety from one cut to the next.

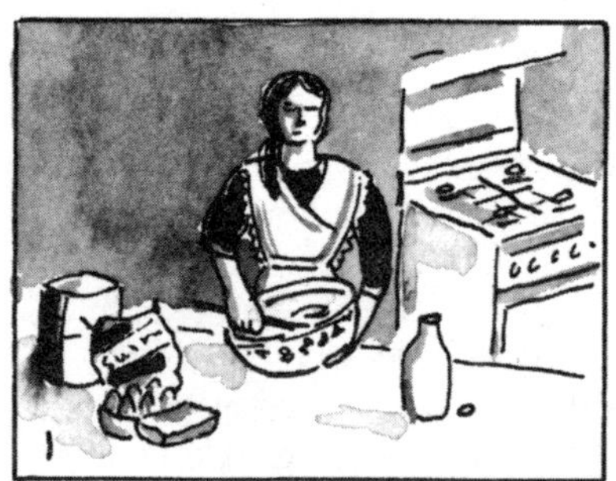

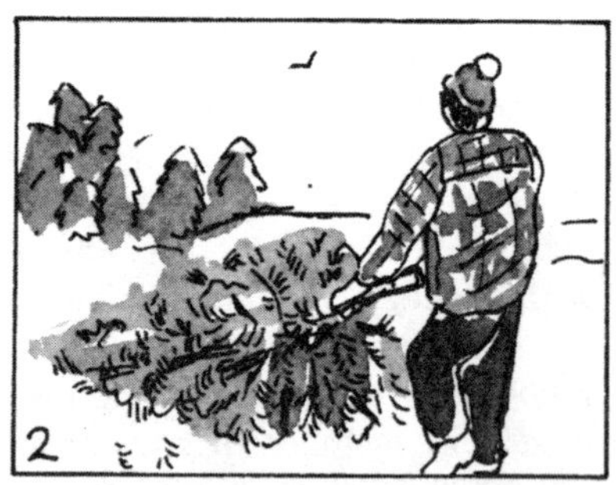

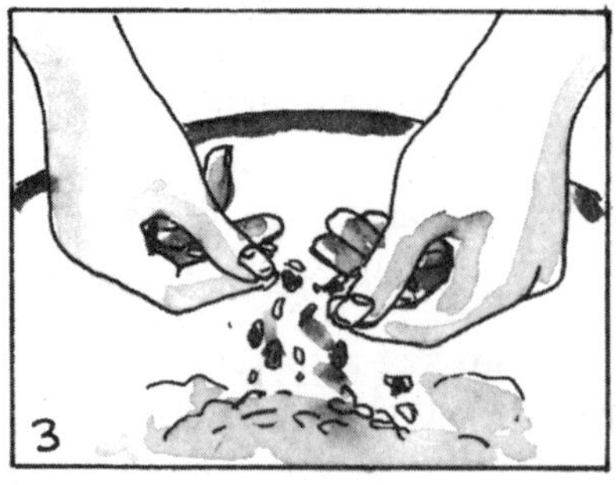

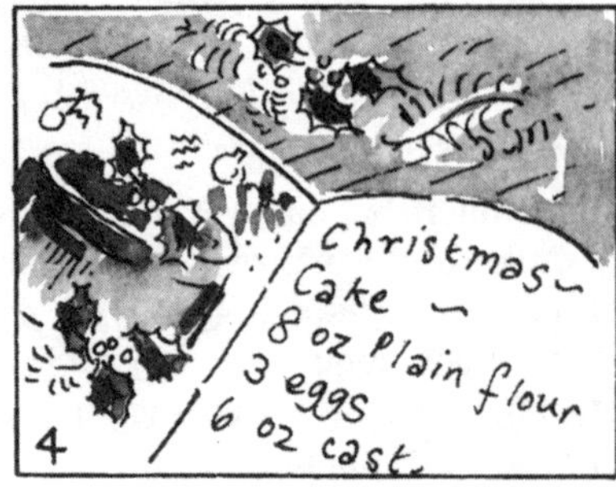

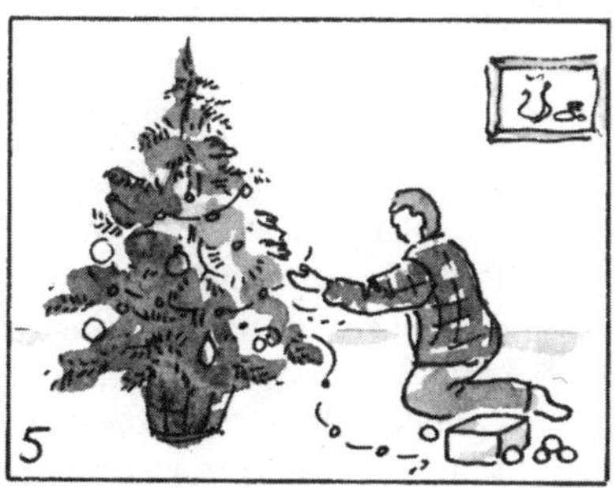

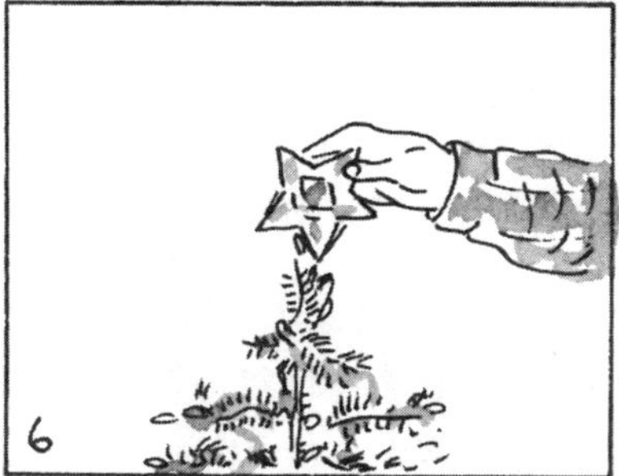

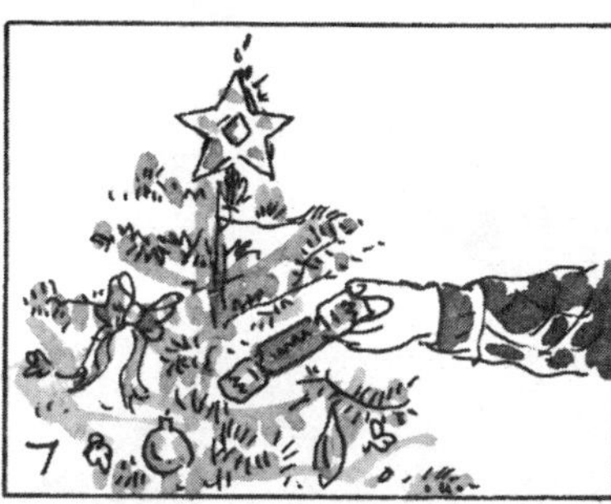

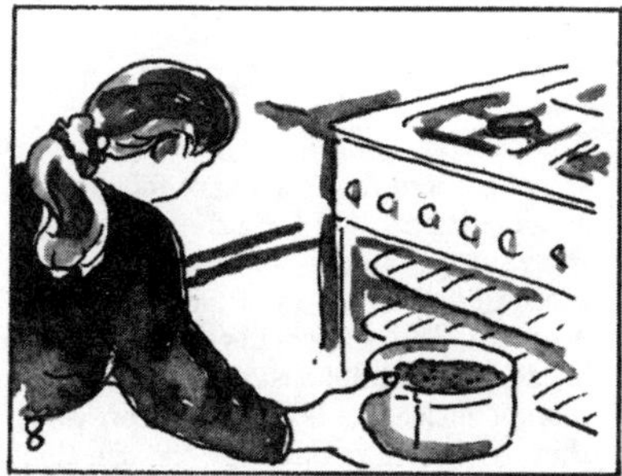

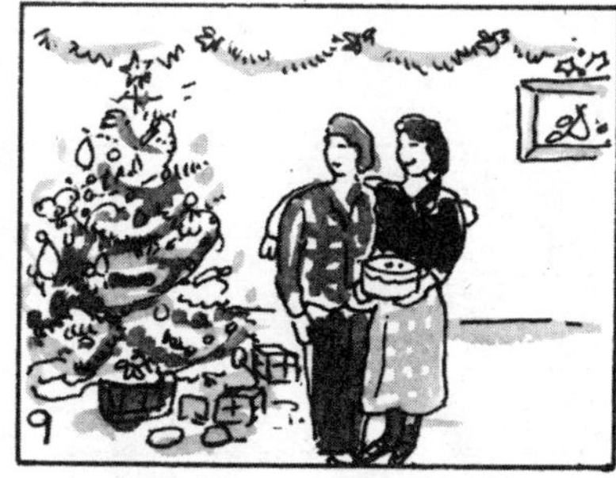

Parallel action allows you to show two different stories at the same time and to suggest they are happening simultaneously and are somehow related. Here the man dresses a Christmas tree, while the woman bakes a cake. It is not until the last shot that we see that both events have taken place in the same house

Illusion

Because we all make certain assumptions when we watch films or videos, it is possible to use the conventions they are based upon to bend the rules. For instance, because we assume that one shot is connected with the next in some way, we can put two completely different things together and make them look as if they are related.

A simple example is to start with a long shot of the outside of the building – and then move to an interior shot. We assume it is the same building. But there is no real reason that it should be. Interior shots of two people talking could be shot anywhere. So your exterior could be of the White House – and so con your viewers that the next shot is taken in a government office. Of course, this technique is commonly used in the movies, where most interior scenes will be shot in a studio. But there is no reason why you should not use similar tricks in your own movies.

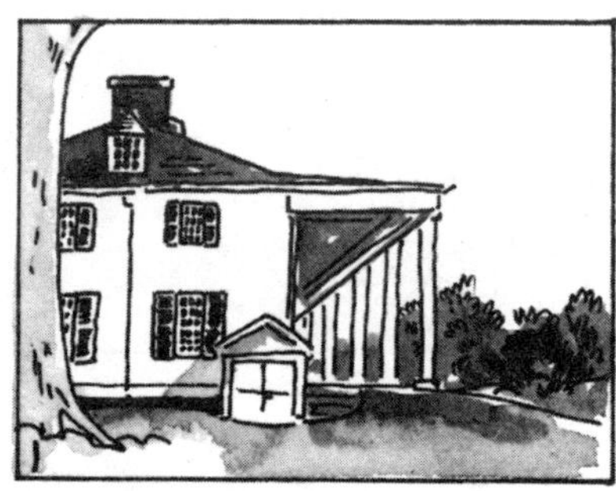

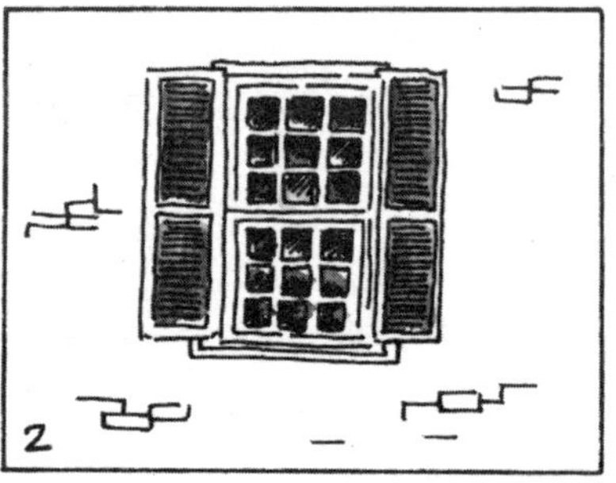

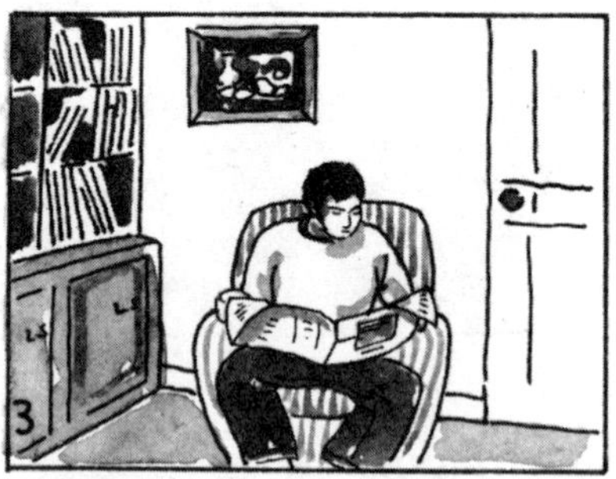

It is easy to fool the viewer with video as we assume that each shot follows on in some logical way from the shot before. Here we see a grand old mansion, followed by a close-up of one of its windows. The next shot is the inside of the house. Our natural assumption is that the room is inside the house in the previous shot. The camera operator can use this assumption to make it seem that the subject is inside the unlikeliest of places . . . The White House . . . Buckingham Palace . . . The Kremlin . . .

A clever use of the same technique might be to have a shot of your children peering through the bushes at the bottom of the garden, and then cut to a close-up shot of a lion roaring. With a suitable reaction from the children, it looks like the children are seeing the lion. Of course, the lion could have been taken on a completely separate occasion at a safari park. You would need careful matching of surroundings, lighting and camera angles to make it look convincing, but even if this is not possible, it might add a touch of comedy to your video.

Another example of how this technique could be used for comedy (or drama) might be to start with a close-up shot of a hand holding a worm. You then cut to a shot of someone putting something in their mouth. It looks like the person has eaten the worm! In truth, the worm never went near anyone's mouth – and the hands in the first shot might even belong to someone else entirely! Again an appropriate reaction from a bystander can help make the trickery look convincing.

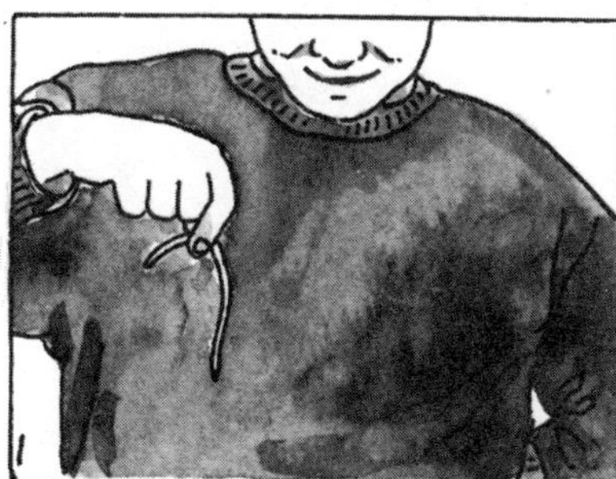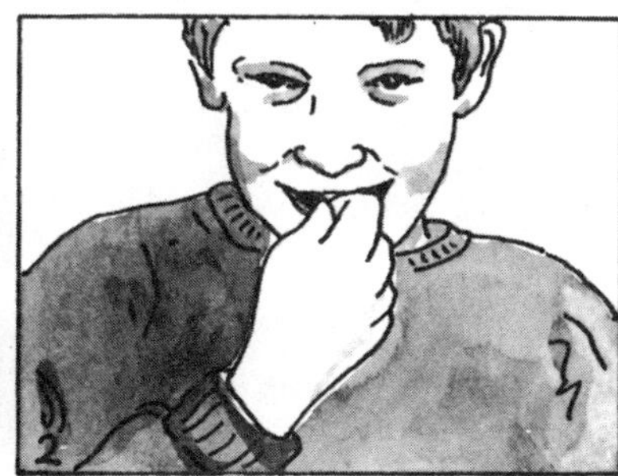

Shot 1 shows a boy holding a worm. Shot 2 shows a boy putting something into his mouth and eating it. By cutting from shot 1 to 2 you imply that the boy is eating the worm, even though you are not actually showing this (or recording it!)

You can use this false implication in almost any video. A spectacular sunset shot on the other side of the world, if cropped sufficiently, could make the ending for any holiday video. The possibilities are only limited by your imagination – and your skill at covering up the truth!

A clean break

There are times when a simple cut between scenes is completely inappropriate – simply because there is no connection between them. A video of a touring holiday would look funny if it suddenly cut from Canterbury Cathedral to Margate pier. Such complete breaks in subject matter are easily possible as you move from one day's shooting to the next. If your video is made up of clips of your children, there may even be weeks from one shot to the next.

What is needed is a complete break between the two different events. One possible ways of doing this is to use the fader that is built into most camcorders. A much stronger method, however, is to record 3–5 seconds of black between the unrelated scenes. This is easily done by recording a brief shot with your lens cap on. As your camcorder will continue recording sound even when there is no light entering the lens, you should do this somewhere quiet. Alternatively an unwired plug put in your camcorder's microphone socket will ensure that nothing can be recorded on the soundtrack. A title or date could be superimposed over the black just before the new sequence starts to help reinforce the fact that this is a new chapter in the video. Remember to establish the new location straight away with the new sequence, so that there is no confusion in the viewer's mind.

12
SOUND

It ought to go without saying that sound is a vital part of video. However, it is all too easy to forget about the soundtrack while using a camcorder, as we automatically tend to concentrate on the picture side of things. Even in our terminology we ignore the audio side – the word video implies pictures alone, even though the medium is an audiovisual one.

Sound is not only recorded automatically by the camcorder, but it is also automatically synchronised with the video. To a large extent, it is possible to concentrate on framing up our shots, and leave the camcorder to record the sound that accompanies them. But there are dangers involved in doing this. The main trouble is that sound does not travel as well as light. We can see for miles into the distance, but we can have difficulty hearing someone talk if they are more than a few yards away. The camcorder has the same problem – only worse. The human ear has the ability to ignore certain sounds that it does not want.

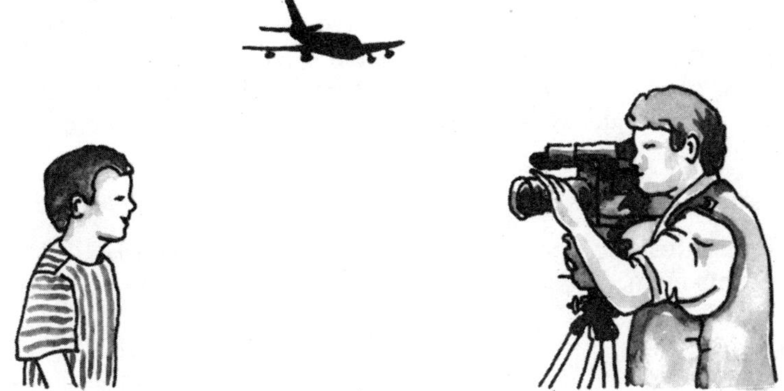

The person behind the camcorder can hear what the person is saying – but the camcorder will not because of the noise of the plane overhead

to hear. You are sitting in a crowded bar, hundreds of people are talking at once, music is blaring out in the background, traffic races past in the street outside, but still you can concentrate on what the person next to you is saying, and can largely ignore the competing cacophony. The camcorder's microphone cannot be selective and therefore it is easy for the sounds that we want to hear to be lost.

What is sound?

All sounds are created by vibration. The simplest sound is caused by hitting an object, causing it to vibrate. As it moves to and fro it sets off a series of waves in the air around it – like dropping a stone in a pool of water. Different sounds can be made by plucking a taut string to make it vibrate, or forcing air through a hollow tube – again causing pressure waves in the surrounding air.

The way we hear these waves is that when they enter the ear they cause the eardrum to vibrate and the cochlea converts these pressure waves into electrical signals, which in turn are interpreted by the brain. A microphone works in a similar way. A thin metal diaphragm inside the microphone is caused to vibrate by the pressure waves in the air and again these vibrations are converted into an electrical signal.

Obviously there are many different sounds that the ear or a microphone has to be able to differentiate. A number of factors are used to make this possible:

Frequency The faster the sound source oscillates, the higher its frequency or pitch. We measure this pitch by the number of oscillations per second (cycles per second or Hertz (Hz)). The human ear is capable of hearing a frequency range of 15-20,000Hz – although this upper limit deteriorates significantly with age.

Frequency ranges

Human hearing	15–20,000Hz
Trombone	100–850Hz
Male voice	80–800Hz
Female voice	200–1,000Hz
Piano	27.5–4,186Hz
Mono linear video track	80–10,000Hz
HiFi soundtrack	20–20,000Hz

Harmonics Although every sound has its own fundamental frequency, a particular note played on a piano will sound very different from one played on a trumpet. The reason is that these notes are not simply made up of the one frequency – each has accompanying 'overtones' of a higher frequency. It is these overtones that give an instrument its own distinctive sound. It is for this reason that a good HiFi recording system has a far wider frequency range than is required to record the basic frequencies of speech, or even a church organ. To be able to identify a sound correctly, the higher overtones have to be recorded as well. The mono linear track with its narrower frequency response cannot record all the accompanying harmonics to a sound and therefore sounds inferior.

Volume The way the human ear tells loudness depends greatly on circumstances, such as how loud the previous sound it heard was, the amount of background noise, etc. It is measured in decibels (dB), and is such that an increase in 3dB equates to a doubling in loudness. Large differences in loudness are hard to record effectively with a camcorder, because the camera is constantly adjusting volume using a system called the Automatic Level Control (ALC). We shall look at the pros and cons of this system later on in this chapter.

Loudness scale

Average house background noise	40dB
Average office background noise	60dB
Conversation (one yard away)	80dB
Moving train (ten yards away)	100dB
Thunder	120dB
Jet aircraft	140dB

Acoustics The quality of sound is affected by the way it bounces off objects before it is recorded. It is for this reason that someone speaking will sound different in a completely empty room from the way they will in one that is fully furnished.

When a sound wave hits a hard surface (such as walls, mirrors, or windows) little of the sound is absorbed – so the reflected sound waves are almost as strong as the original. Higher frequency waves can even be stronger. On the other hand, when a sound wave hits a soft surface (such as curtains, carpets, and sofas) some of the energy in the wave is absorbed. Not only is the reflected wave weaker, but higher frequency waves are absorbed more readily – so are missing from the reflection. The resulting sound is more mellow and muted.

If a location where you are recording is full of hard surfaces (such as a church, garage or bathroom) the sound is repeatedly reflected from one surface to the next and the original and reflected sounds become so intermixed that they are recorded on top of each other. This is known as an acoustically 'live' environment. The opposite is a 'dead' environment, where a room is full of soft surfaces. Here the original sound and the reflections are quickly muffled.

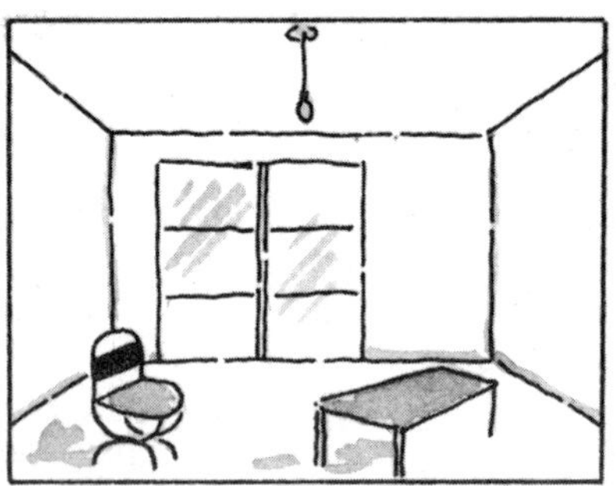

Live acoustic: the more empty and unfurnished a room, the more the surfaces reflect sound

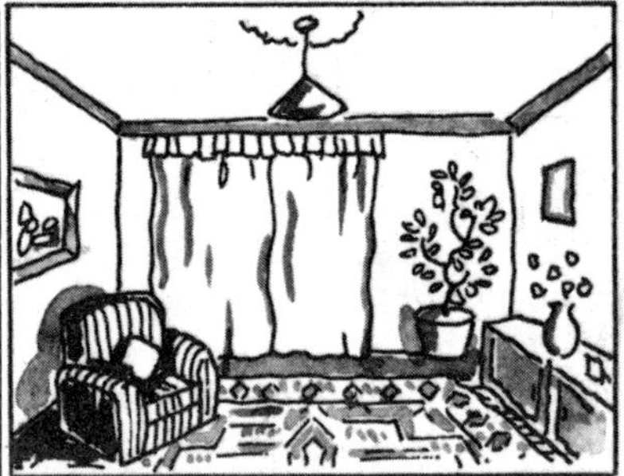

Dead acoustic: furniture, carpets, and curtains all help to absorb the sound

The acoustics of a particular location will have an effect on where you will want to place the microphone to get the best recording. In a live acoustic you will want the mic close the subject to drown out the effect that the acoustics will have. You will want to use the same technique in a very dead environment too, such as recording outdoors when there are no surfaces for the sound to bounce off at all. A living room, full of soft furnishings, will be ideal for recording speech if you want to hear every word clearly.

The acoustics of a room can be changed to some extent to suit your purposes – by removing or adding a rug on the floor, or opening or closing the curtains.

Automatic level control

In a professional film crew there is one person whose sole responsibility is to check that the sound is recorded correctly. This person constantly monitors the sound, altering the level it is being recorded at so that it sounds right on tape. The camcorder operator does not have this luxury. You are a team of one and there is enough to do without having constantly to change the sound recording level controls. For this reason, every camcorder adjusts the recording levels automatically. The circuitry used is called the Automatic Level Control (ALC). This ensures that the volume of your soundtrack remains relatively even while you are recording.

If you did not have this ALC circuitry the sound would constantly change in volume. A person talking three yards away may sound fine, but if he took a couple of steps forward, it would sound as though he were shouting. If he moved back, you would hardly hear him at all on the recording. Such a state of affairs would clearly be unworkable.

However, the ALC does have its shortcomings. In particular, it cannot distinguish the occasions when you want a noticeable increase in volume on the soundtrack. For instance, if you pop a balloon – the ALC will drop the volume down, and the shock effect of the bang will be completely lost. In quiet situations, such as an empty room, the level might be boosted to such an extent that you can hear the camcorder's motors and even your own breathing. The trickling of a mountain stream can sound the same as a mighty waterfall. Unfortunately, very few camcorders have a manual recording level control to allow you to correct for these types of occasion.

The built-in microphone

The microphone that comes built into your camcorder, as we saw in an earlier chapter, is a jack-of-all-trades. Its key problem is that it is stuck on the camcorder. For many shots, therefore, it is going to be a fair distance away from the sound source. Sound does not travel well and we are unlikely to be able to record exactly what we want to hear. The ALC will not help here – as it will adjust to the volume of the background noise, and not to the voice in the distance.

Your built-in microphone will be one of three types:

Omnidirectional That is it will pick up sound from all around – in a 360 degree circle. Sounds from behind the camera, therefore, will be recorded as well as those in front. With this sort of microphone you will have to be particularly careful of noises off-screen – people talking to the side of the camera or just behind will sound loud if they are close enough, for instance.

Unidirectional It will pick up sound primarily from a forward direction, from a 200 degree arc. Although sound from behind the camera is less of a problem, it is still easy to record unwanted off-screen noises.

Zoom Found on some top-of-the-range models. This allows the microphone to change its direction angle as the camcorder zoom's focal length is changed. So with a wideangle lens it will accept sound from a wider arc than with a telephoto lens. However, the sound arc will always be significantly wider than what is seen on screen, so the mic will not be able to zoom in on a voice in the crowd. Some will give you manual control over the mic's directionality.

Monitoring the sound

One of the most useful accessories that you can buy is a set of headphones. These allow you to hear exactly what you are recording – so at least you are aware of any problems with the soundtrack as they happen. Unfortunately, many camcorders do not have the requisite headphone socket to allow you to do this.

If you can, a pair of headphones will allow you to spot when off-screen noise becomes a problem, or when the ALC is stopping you from getting the effect that you want. Any sort of headphones is better than none – even the earplug-type that are used with portable cassette players of the Walkman variety. Bigger headsets that completely enclose your ear are best, as they cut out most of the ambient noise, so you can concentrate on exactly what the camcorder is picking up.

Making the most of a built-in mic

The key thing to remember when using your camcorder's on-board mic is never to shoot from further away from your subject than is absolutely necessary. Move in close rather than use your zoom lens. This is particularly important if you are recording speech. If you are recording people talking there is normally no reason why you should not get within two or three feet of the speaker – and still get a range of different shot sizes, if necessary.

For many shots it may not be so important where you record the sound from – the ambient sound of a garden is much the same from one end to the next – so you can shoot from any position and use your zoom with impunity.

One of the greatest problems you will come across, whatever mic you are using, is unwanted background noise. Wind is undoubtedly the worst offender. One answer is to try and find yourself a spot where you are more sheltered – as the wind tends to whistle through the microphone housing itself, making it sound much worse than it actually is and blocking out all other sounds. Find a doorway to stand in, crouch behind a wall, stand with your back to the wind, or get someone to stand upwind of you – giving the microphone that extra bit of shelter.

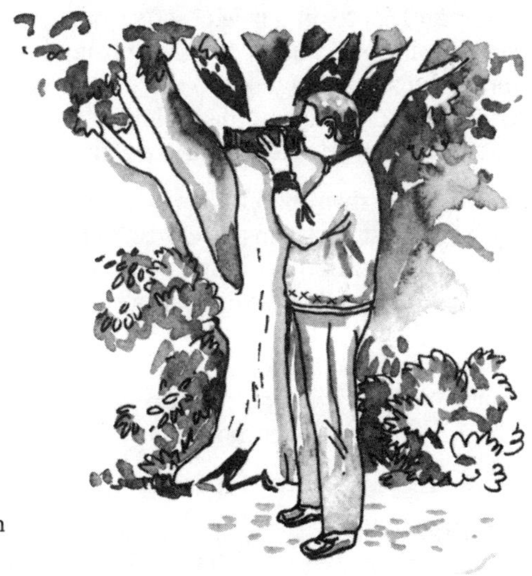

Find a sheltered spot out of the wind to shoot from to cut down on unwanted background noise

Some microphones have a built-in wind filter – which electrically cuts out the offending frequencies when activated. Alternatively, you can

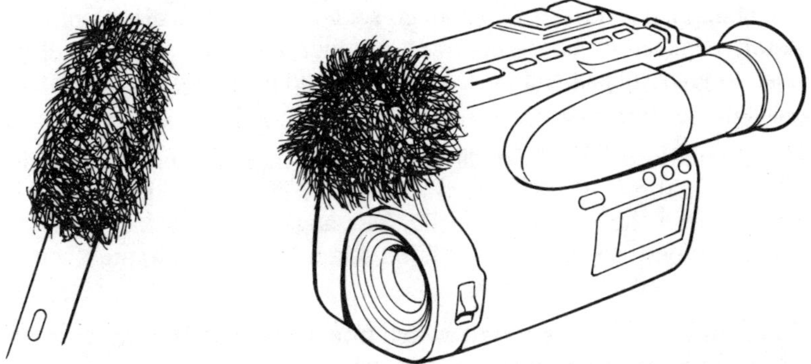

Two types of windshield: (a) for covering stick microphones; and (b) for built-in microphones on palmcorders

get accessory windshields that fit over your microphone, and are designed to muffle the unwanted sound.

Add-on microphones

The best solution to a whole variety of problems involved with sound recording is to fit an accessory microphone which is designed, and specifically positioned, for the job in hand. As soon as the accessory microphone is plugged into the camcorder, the built-in mic is disabled. Unfortunately, again, this solution is unavailable to many camcorder owners, as their machines are not fitted with the necessary mic socket.

The add-on microphone has two distinct advantages over the built-in variety. First, you can place it close to the sound source, without having to shoot from that position. You can therefore shoot a conversation as a long shot, without worrying about whether you will be able to hear what is being said on the recording. Secondly, different microphones have different characteristics – some pick up sound from all around, while others are highly directional, thereby excluding much of the unwanted background noise.

Here are some of the types of microphone that you may come across:

Cardioid The basic unidirectional mic, accepting sound in a 200° arc. It gets its name as its directionality forms a heart- shaped pattern around the mic. It is good for recording speech at close distances (up to about three yards).

Supercardioid Another unidirectional mic, with a narrower angle of acceptance (around 120°) – although sound outside this arc is not completely eliminated. These mics are designed for when you can not get close to your subject and therefore are forced to record the sound from a distance.

Hypercardioid The most directional type of microphone, accepting sound only through a 90° arc. Useful for wildlife – although it must be remembered that this is a much wider angle of acceptance than any telephoto lens, so care must still be taken to avoid unwanted sounds being recorded.

Zoom mic This allows you to switch from a cardioid to a supercardioid acceptance pattern at a flick of a switch.

Tie-clip mic A very small microphone which is designed to be worn

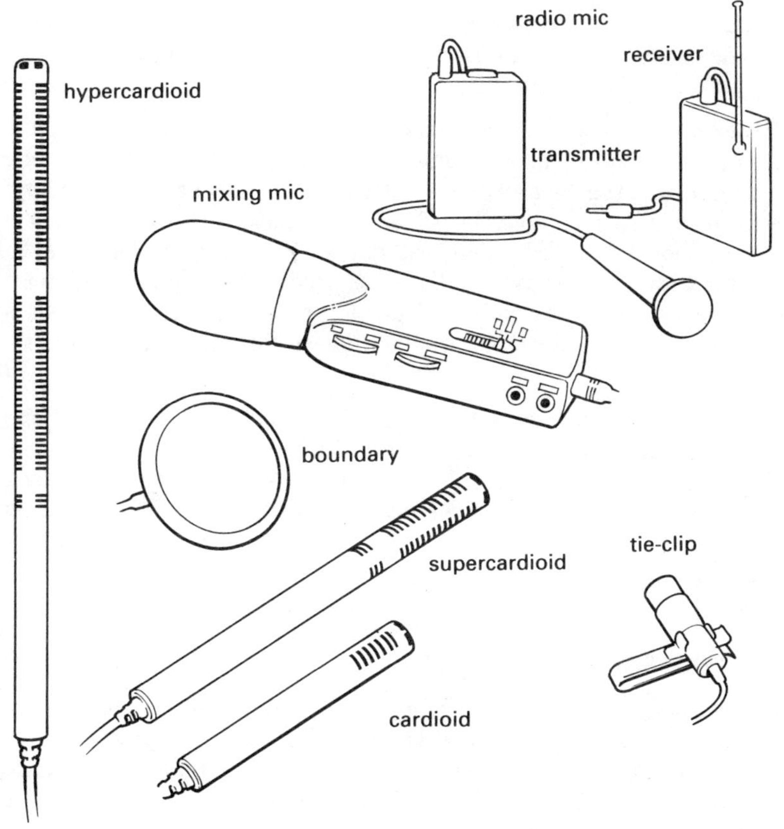

by a speaker without being too noticeable in the shot. Such mics are generally omnidirectional, but as they are within a few inches of the speaker's mouth, there is little problem with background sound (other than avoiding clothes brushing against the mic itself).

Narration mic An ice-cream cone shaped omnidirectional microphone for recording speech, when the speaker's lips are just a couple of inches from the mic. Ideal for when you want the microphone in shot for a newsy feel to your video. Also used for adding commentaries in post-production.

Boom headset A pair of headphones with a microphone mounted on a boom so that it can be positioned just in front of the lips – as used by telephone operators. These are mainly used in post-production when mixing sound as you add a commentary.

Boundary mic An omnidirectional microphone for recording speech. It is designed to be placed flat on a table or a stage, and so that it does not pick up echo – eliminating the effects of a live acoustic.

Mixer mics These combine a microphone with a miniature audio mixer. This allows you to plug a separate sound source, such a portable tape recorder, into the microphone – so that you can mix background music with the live sound as you record. Ideal for those who want to produce a professional sounding video, without the trouble of having to fiddle around with sound mixing at the post production stage. These mics may also provide you with an input for a second microphone, or a headphone socket (for those who don't have this facility on their camcorder).

Radio mics These convert the electrical signal from the microphone to a radio signal which is then converted back to an electrical signal by a receiver unit at the camcorder. Ideal for picking up sounds over long distances (up to 200 yards). The microphones are usually of the tie-clip or cardioid variety. You should ensure that any transmitter you buy uses an approved radio frequency.

Stereo sound

An increasing number of camcorders are now available that record stereo HiFi sound. By recording two different soundtracks simultaneously, from slightly different positions, you add a feel of distance to a soundtrack. People who move across the frame also sound as if they are moving across the frame.

We have all got used to the spectacular effects and added depth that stereo sound affords through recorded music. Television audiences are gradually getting used to this phenomenon too, as they upgrade their TV receivers.

Recording sound in stereo brings its own problems. A built-in microphone can only give a certain amount of stereo effect as the two parts of the microphone are so close together. Increasing the stereo effect means placing two microphones further apart, but in most situations this is impractical for the camcorder user on the move. Obviously the whole variety of different microphones listed above are available in stereo versions – and these will generally give a better stereo effect than the built-in mic will allow.

If you have a stereo camcorder, it makes sense to try to get the most out of the twin soundtracks. The best advice is to make sure that you listen to your recordings in stereo when you play them back. If you have a stereo television this is straightforward. Even if you do not, you can get an even better stereo effect by rooting your audio out signals from your camcorder through your HiFi system. The connections are reasonably simple – although it might mean extensive wiring if your HiFi is a long way away from your television. Instead of connecting the audio leads from your camcorder to the television, you connect them to an auxiliary inputs on your amplifier. The video signal is connected direct to the television as usual. Ideally, your speakers should then be placed either side of your television to get the most realistic stereo effect.

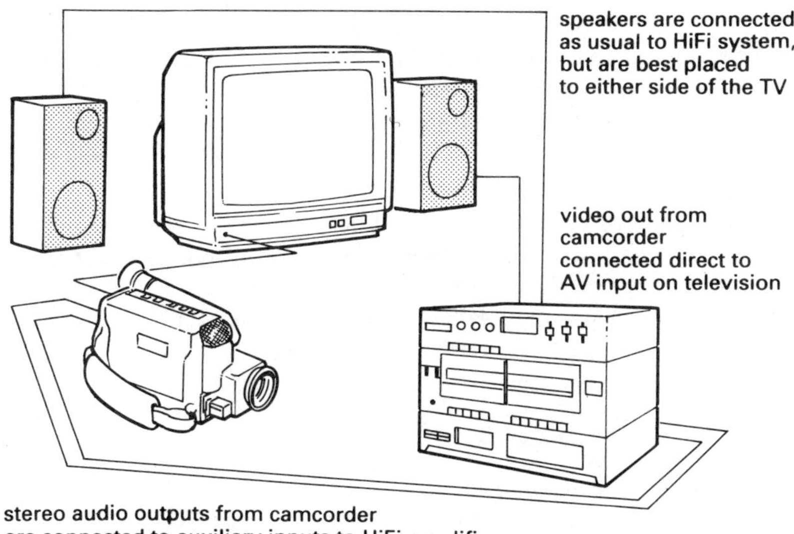

Set-up for using a HiFi system for listening to the soundtracks on a stereo camcorder

Cutting between shots

As explained in Chapter Eleven, the problem with video is that the sound is not continuous. Each time you stop recording to change shots, there is a break in the soundtrack. On some occasions, the difference in the sound from the end of one shot to the beginning of the next is insignificant. Background noise can be cut together with little worries about there being an audible jump when it is replayed. However there are occasions when you need to time your cuts so that they coincide with a natural break in the soundtrack:

Speech. If someone is speaking, wait until the end of a sentence before ending the shot. A recorded conversation made up of individual sentences will generally sound okay, even if much of the conversation is cut out. If this is going to be a problem, and the speech is secondary to the speakers, shoot it in such a way that the words are not audible – shoot from further away with a telephoto lens setting, for instance. Then you can cut when you want.

Music. This can be a particular problem if you are shooting somewhere where the music is playing continuously. Any break will be very noticeable. Again try shooting so that the music is only just audible in the background. If you have a VHS-type camcorder with an insert edit facility you may be able to record a single piece of music in its entirety – then insert edit your other shots over this. The drawback is that you will have to use the sound on the mono linear track alone, as the 'live' sound on the HiFi track (if you have one) will have the same jump problem.

Planes, trains and automobiles. Short-lasting loud noises can be another problem when trying to end your shot. A sound of a passing train in one shot which suddenly disappears in the next will sound suspicious. The only realistic way to avoid this problem is to shoot the first shot again when the noise has passed.

13
ADVANCED PRODUCTION

Preparation is the key to a professional video production. While an amateur may be happy to shoot a video of an event as it happens, a pro will try to avoid this off-the-cuff approach. The less things are left to chance, the better. Every detail should be planned beforehand. In fact, it is not uncommon for the videomaker to know exactly what his production will look like even before a single shot is recorded. By visualising all the shots before starting the videoing, potential problems, such as jump cuts or breaks in continuity, can be avoided. This approach is not just used for dramas – it can be used equally successfully for documentaries or covering live events. It does not mean that unplanned happenings have to be ignored. But at least spur-of-the-moment changes can be incorporated into an existing framework.

Of course, the advanced production process relies heavily on the fact that the raw footage will be edited later on. But even if you are planning to chop out parts of the video at the editing stage, you should still try to shoot little more than you actually need. In this way the post-production stage is just a tidying up operation and not a wholesale rebuilding job.

Storyboards

A popular way of mapping out what a video is going to look like before you start is to use a storyboard. In essence, this is a set of simple drawings with each sketch representing a single shot in the video. You don't need to be a skilled artist to make these drawings; stick men are perfectly acceptable. After all, it is for your use only. As long as you can understand what the squiggles signify, it doesn't matter how simplistic the drawings are. Each drawing is usually accompanied with a short caption that describes each shot. This would usually include:

- The shot size (MS, LS, CU, etc.)
- The length of the shot in seconds
- Any camera movement to be used with the shot
- Description of what is in the picture
- A note of any sound that must be recorded with the picture
- A shot number – so you can see which drawing follows which in the storyboard.

A storyboard will allow you to see any pitfalls in the production. For example, it might show a break in the continuity of line – with a person looking to the right in one shot, and to the left in the next shot. It will also help you to find places where you can use more interesting shots – such as a low-camera angle, or limited depth-of-field. Finally, the exercise will also help you to keep your videos short – you will be able to see at a glance which shots are really necessary to tell your story.

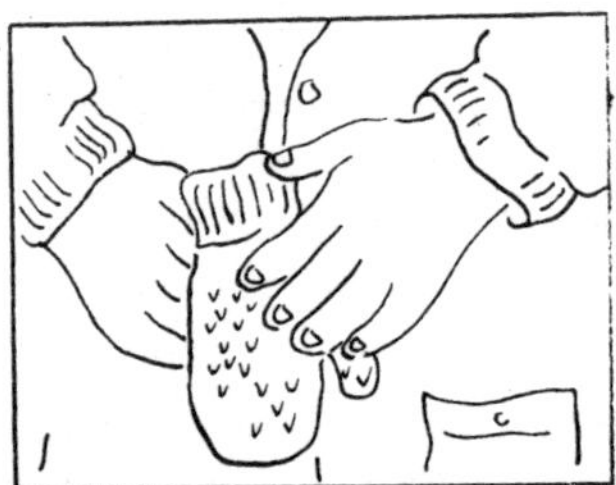

BCU. Boy putting on gloves.
6 secs

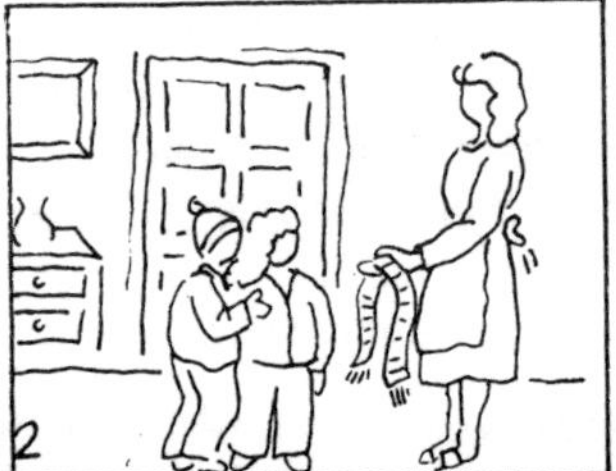

LS. Jack and Jill by back door with Mum. Mum gives Jill scarf. 8 secs

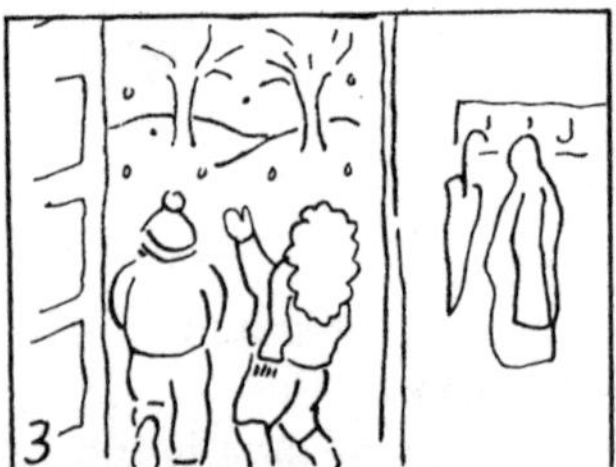

LS. Jill and Jack run into garden. Camera tracks through door to garden covered in snow. 12 secs.

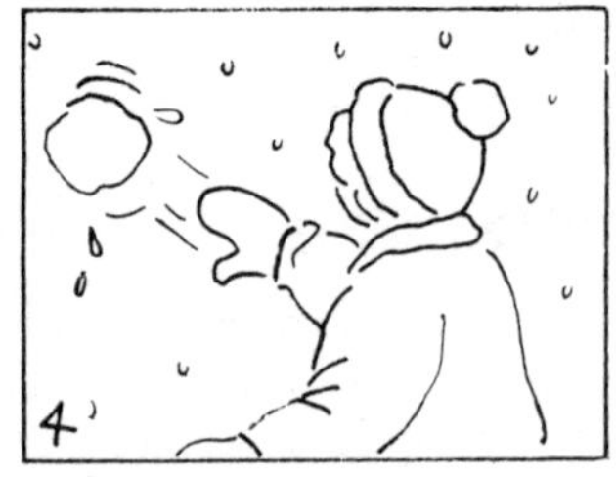

MS. Jack throws snowball. 3 secs.

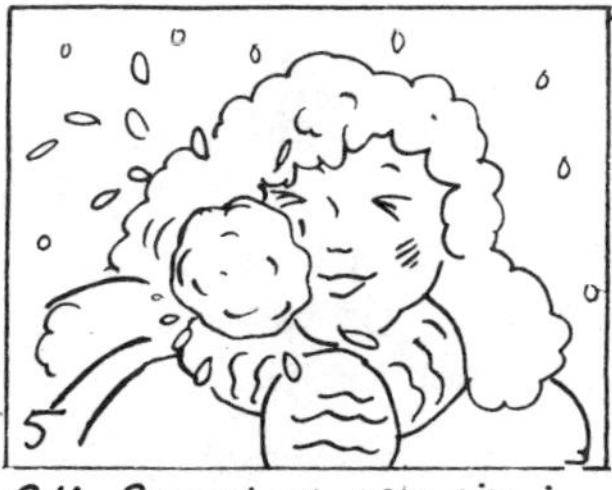

CU. Snowball hits Jill in face. Jill cries out. 5 secs.

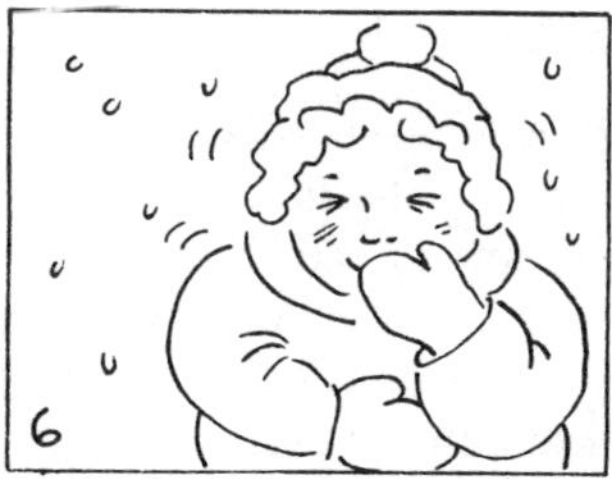

MS. Reaction shot of Jack laughing. 2 secs.

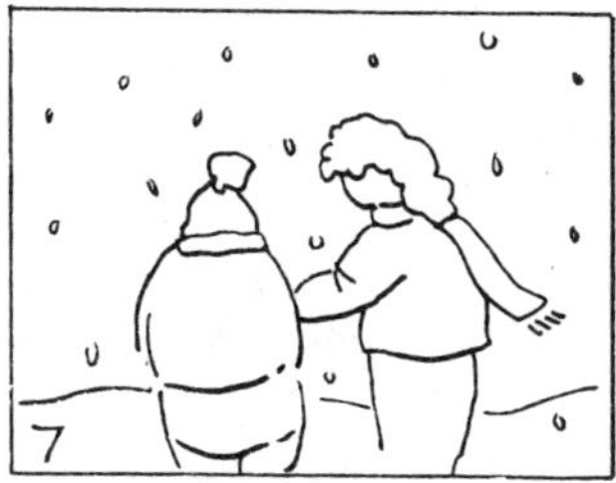

MS. Jack and Jill, backs to camera, working at something camera can't see. 5 secs.

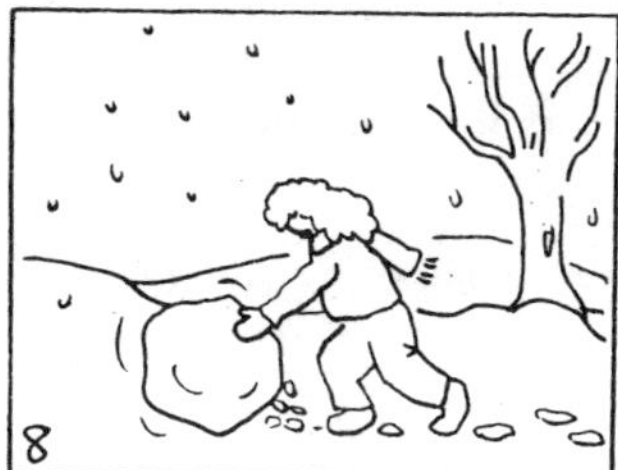

LS. Jill rolling big snow ball. 5 secs.

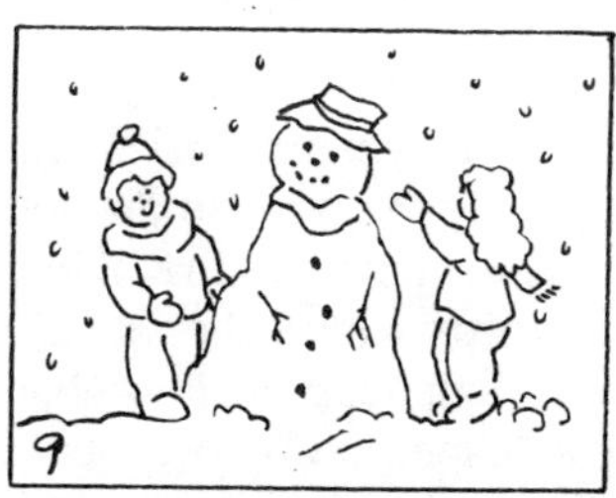

LS. Jack and Jill putting finishing touches to snowman. Jill puts scarf round snowman. Jack puts hat on its head. 12 secs.

Typical storyboard – the drawings can be as rough and ready as you like, they only act as a visual planner and reminder when shooting

Rather than having to storyboard all of your video, you may find that the technique is most useful for planning key sequences such as the introduction or conclusion.

Scripting

In the movie business, nearly every film begins with a script. In it the full dialogue of the film is accompanied with the briefest of description about the settings and the action. It is only later that the film is fleshed out by the director into the full storyboard. Unless you are making a drama, you are unlikely to write a script for a video before you write the storyboard. In fact, you may not need a script at all if the words are secondary to the pictures.

A script is useful, however, if you are shooting an extended piece of dialogue, such as the commentary for a holiday video. It is particularly important because it is easy for commentaries to state the obvious. The pictures tell much of the story on their own – so the words must add something new, and not just describe what is self-evident from the pictures.

If you were shooting a sequence of shots of Buckingham Palace, for instance, it would be unnecessary to say 'Next we went to Buckingham Palace. It must have hundreds of rooms. Soldiers keep a constant guard outside.' These things should be obvious from your pictures. It would be more informative to say 'Queen Victoria was the first British monarch to live at Buckingham Palace – and it his her memorial that you see first as you walk up The Mall'. For a more lighthearted commentary you might say 'We knew the Queen was at home as the Royal Standard was raised high above Buckingham Palace – but she didn't come out to say hello.'

To avoid the problem of stating the obvious, commentaries are usually recorded after the pictures are shot, during the post-production. Commentaries should also be kept short. It is important that they are not continuous; there should be plenty of gaps where the video and live sound is uninterrupted. When writing a commentary allow about three words per second of video. If you are speaking into a microphone, make sure that it is just a couple of inches away from your mouth and speak over it rather than directly into it. If you can, choose somewhere with a dead acoustic so that your words will sound clearer.

Staging the action

If you are shooting a drama, it is easy to get your actors to repeat a scene if you were not quite happy about the framing or sound. It might seem impossible to exercise this control over your subjects when shoot-

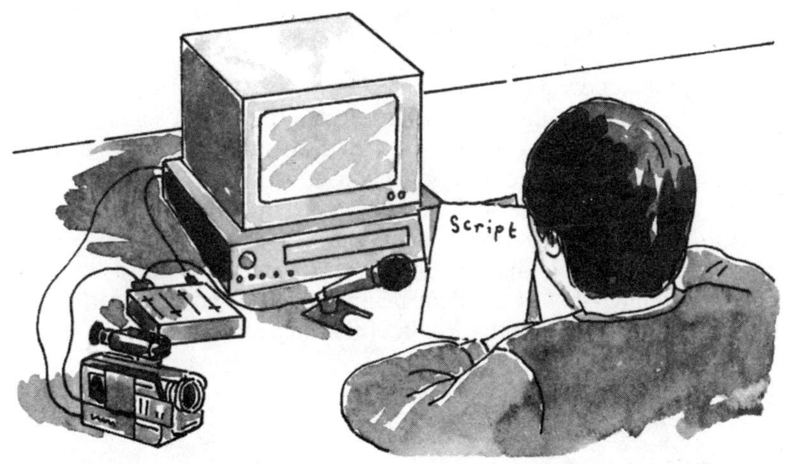

Commentary is usually mixed in after shooting, at the editing stage. Write out what you are going to say before you start. Keep the microphone a couple of inches from you mouth and speak over it, rather than into it

ing documentaries or live action, but it is perfectly possible to restage events. Some might call it cheating, but if you saw a TV documentary crew in action, you would be amazed at just how many shots are set up specially, or repeated, for the benefit of the camera. The point is that the videomaker cannot always predict what is going to happen next. If he guesses wrongly, the shot may look less than perfect or it might be missed completely.

What is more, although it might be possible to get a single shot of something as it happens, it is much harder to get an effective sequence without missing out huge chunks of what is going on. A good example is a video of a car journey. If the camera is travelling in the car, it is impossible to get shots of the car arriving at its destination by recording things in the normal way. You have to cheat. The car has to arrive once, drop off the camera operator, turn round, drive away and then arrive again – this time so that it is captured on video. The technique is used so often on television that no one questions how the camera has managed to keep one step ahead of the action.

Similarly, it is easy to forget to take a relevant shot as it happens. At a birthday party, you may not have the opportunity at the time to take close-up shots of the presents being opened – or to take reaction shots of the child's face. By restaging this small event some time later you have all the time you need to get the necessary shots.

Repeating the action in this way is not only useful for missed shots, or for shots that would otherwise be impossible, but it can also be useful for varying your camera angles through a sequence. For instance, if you are shooting someone entering a house, you can't shoot a close-up of the key going into the lock and also take a long shot of the door opening, without engineering events. But if you do break the flow of things as they happen, there is no reason why you shouldn't get your close-up and long-shot of the door being opened – and even get a shot of the door opening from the inside of the house!

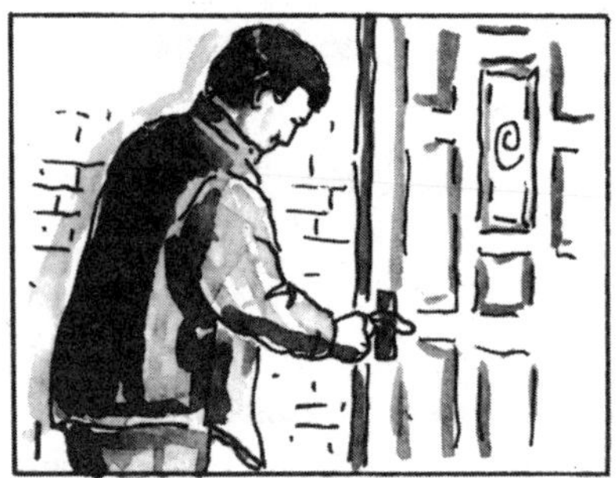

Long shot – man at front door.

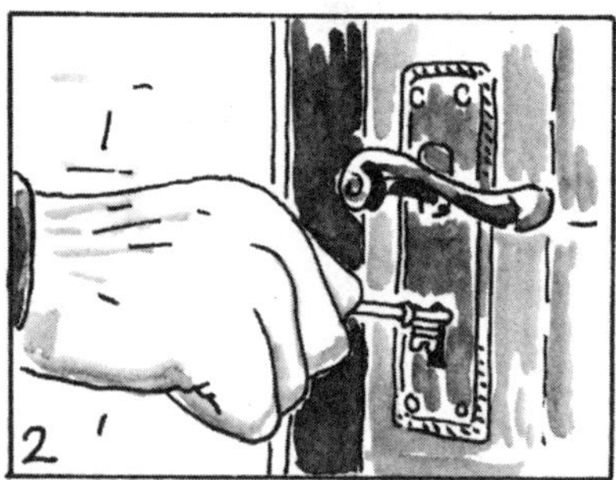

Big close-up – key goes into lock

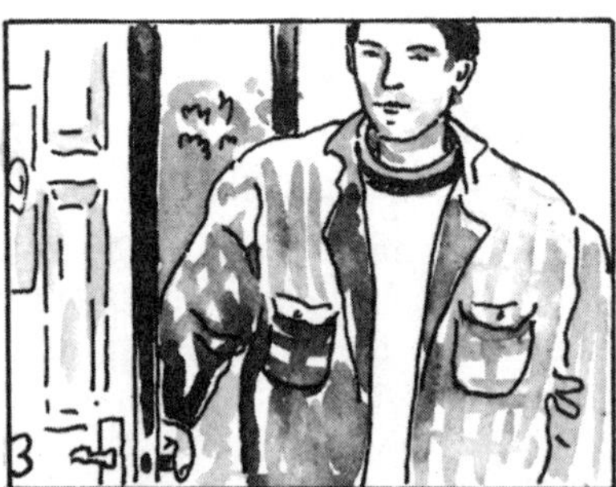

Long shot – man enters house, as seen
from inside.

It would be impossible to shoot these three shots as they happen. By the time the camera person had moved in for the close-up the man would be in the house. Shot C would be even more difficult, as it would mean overtaking the man at the door! The only answer is to stage the last two shots with the man's cooperation

Staging the action in this way demands commitment from the camcorder user. You not only have to get your subjects to cooperate with you, but you must also plan ahead to know when extra, staged shots will be useful. The technique takes time and patience, but the rewards are a video that flows so smoothly from one scene to the next that your viewers are swept along.

Interviews

The interview is a classic example of where shooting the action as it happens means a dull video. Without missing some of the words, it is almost impossible to cut from different shots as the interview is going on. You are therefore forced to shoot the whole question and answer session from the same position. The professional solution to this prob-

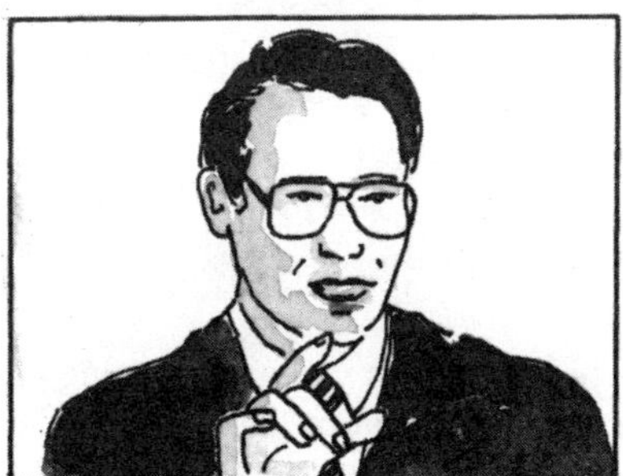

main shot of the interviewee

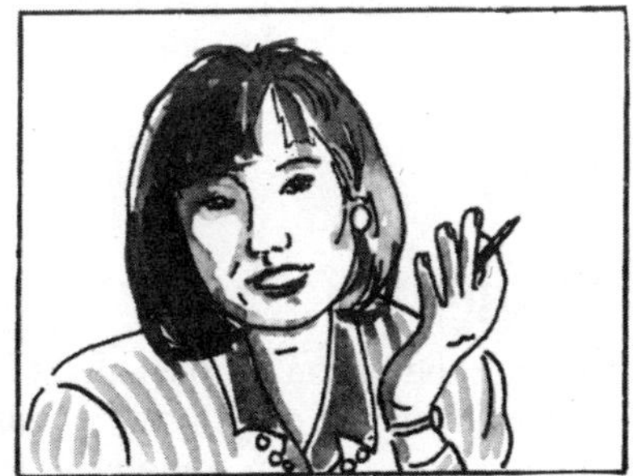

reverse angle shot of the interviewer—the 'noddy'

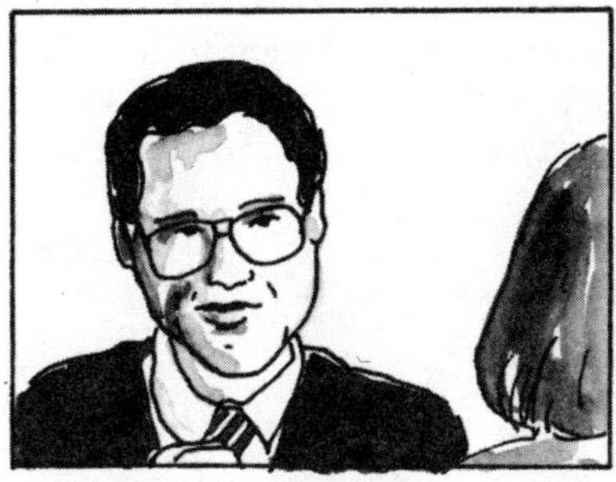

over-the-shoulder shot

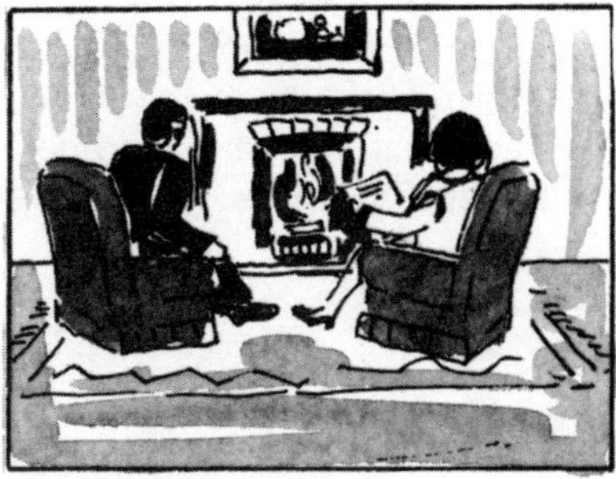

extreme long shot

Some of the different shots used to add shot variation to an interview

lem is to shoot other shots which can be inserted into the main sequence after the interview is over. The normal 'insert' shots are:

- Reverse angle shots of the interviewer – showing him or her intently listening to what the interviewee has to say. These are known as 'noddy shots' in the trade – as they frequently show the interviewer nodding in agreement with what the interviewee is saying. Can be shot after the interviewee has left.
- Over-the-shoulder shots which show both interviewer and interviewee in the same shot. It is taken from behind one of the speakers, so you can see the back of the head of one of them, and the face of the other. This is then inserted when the person with his/her back to the camera is speaking – so that the lips don't give the game away.
- Extreme long shot of the two talking together. The secret here is that the faces of the speakers are so small on the screen that you cannot see what they are saying – so that this shot can be inserted into any part of the interview, without fear that the lips will be out of sync.

Continuity

Staging the action is not without its problems. As you are shooting out of sequence it is easy for two shots that are meant to follow each other to be mismatched in some way. Continuity of appearance is usually the culprit. Your subject wears a jacket in one shot, for instance, but in the close-ups it has gone. On screen you get a noticeable jump between the two sets of shots. But even a change in the lighting can make it obvious that two sets of shots you want to cut together were shot at different times. The only answer is to pay strict attention so that shots that are meant to match fit together perfectly. It is a good idea to review what you have recorded already if you are in any doubt as to what (and what was not) in the previous shots.

14
— SPECIAL EFFECTS —
AND TECHNIQUES

As with any craft, video-making has a whole range of tricks and special-ist techniques for use in specific situations. In fact, in the world of television and film, special effects have grown into their own industry. Most of these techniques are possible with your camcorder – though many demand a huge investment in time and money. In this chapter we will look at some of the more useful and accessible of these tricks of the trade.

Cine to video

Most people have a stash of old cine films or 35mm slides hidden away somewhere in the house which are rarely, if ever, seen. The palaver in setting up the projector sees to that. Transferring this old photographic material to video ensures that it is much easier to view in the future – ensuring that those old memories are seen by a new generation of the family. At the same time, the material can be edited and can be given a new lease of life with the aid of background music or a commentary.

In essence, the technique for copying cine or slide film is simple. All you do is project the images and record the pictures on screen with your camcorder. Although there are special telecine converter units available, they are not necessary for acceptable results. Set the screen and projector up as normal in a darkened room. Then position the camcorder so that it is as close to 90 degrees to the screen as possible, to avoid any distortion of the picture. If possible, set your white balance control to tungsten (to match the projector bulb), and set the focusing to manual (and focus on the screen). To be able to check the quality of your recordings as you go, it is helpful to connect a TV set to your camcorder.

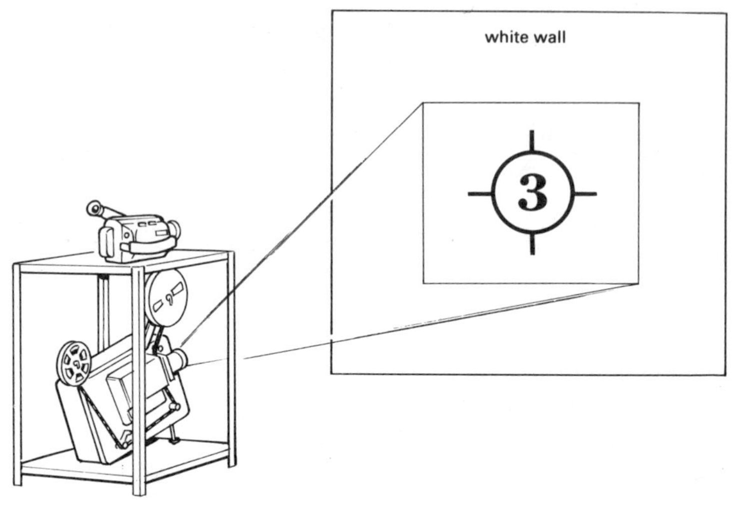

Simple cine-to-video set-up

As most old cine films (and all slides) are silent, you may wish to add a soundtrack – music usually works best, but a commentary might be useful if your viewers are not going to recognise what is in the pictures. As you will have your hands full changing spools or slides, it is best to do this after you have recorded the video. This can be done by audio dubbing the sound through the camcorder's microphone socket (using a microphone mixer if necessary). If your camcorder does not have an audio dub (or mic input) facility, add your music or words while copying the camcorder footage to VHS (see the section on sound mixing in Chapter 15).

One of the problems with transferring cine is that the frame rate of the film is different from that of video – so you can get a flickering effect on screen. This can be eliminated to a certain extent if your projector has a speed adjustment.

With slides, the main problem is that the proportions of the picture are different: 35mm film has a 3:2 ratio, while TV has a 4:3 ratio. You will therefore either lose some detail from the sides of your slides, or the pictures will not fit the TV screen. Upright slides will cause even more of a problem on your landscape-format television. However, you can re-crop each picture as you wish – you can even pan or zoom while recording. You will need to stop and start recording with the camcorder for each individual slide – you cannot let the slide projector run through automatically, as a flash of light will appear on screen between each slide. Vary the amount of time you record each image from 2–5secs, depending on how much information there is in each shot.

As mentioned earlier, there are a wide variety of cine/slide-to-video units on sale. The main advantage these offer is that by using a mirror mounted in a box, or frame, it is possible to position the projector at 90° to the camcorder. This makes straightening up the picture easier and means the whole transfer operation can be done on top of a table.

Rostrum work

Rostrum video is the glorified name given to shooting any piece of flat artwork such as a painting, map or photograph. The key thing to remember here is to avoid any reflections from the surface of the picture. Try to avoid using direct lighting. Sunlight works well indoors if it is not shining directly through the window (so you are using soft, reflected light). Fix the photograph, say, to a wall or a table, and shoot it with your camcorder firmly mounted on a tripod. If you need to use artificial lighting, use two lamps positioned each side of the subject so that they shine at 45 degree to the surface – this will avoid any refec-tions. With large, detailed paintings, or awkwardly shaped works, it is a good idea to show the picture with a sequence of close-ups or by moving the camera while recording.

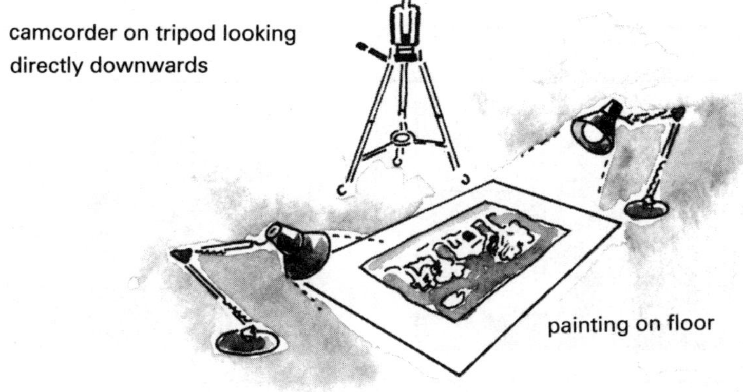

Simple set-up for copying artwork

Animation

As we saw in Chapter 4, a few camcorders are fitted with an animation facility that allows you to shoot just a few frames at a time. Although this will not give you the smooth, seamless effect that professionals get using single-frame film cameras, it can give reasonable-looking animation. How jerky the effect will be will depend on the camcorder – a few record just a couple of frames at a time, while others will record up to eight frames.

Animation involving Plasticine models or drawings depends more upon your artistic ability than your camcorder technique, but whatever your skill in these areas you should start out with a simple project. An easy first stage is to build something piece-by-piece – such as a jigsaw, or a Lego building. Take your short shot after you have put each piece together. When the tape is played back the jigsaw or building will seem to make itself. It goes without saying that the pieces that have been videoed already should be anchored so they do not move between shots

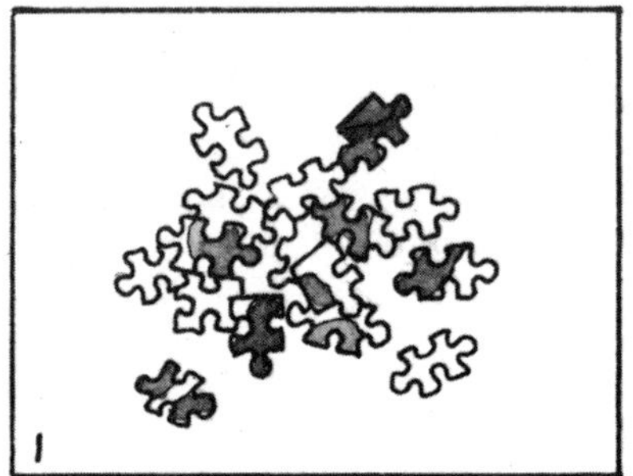
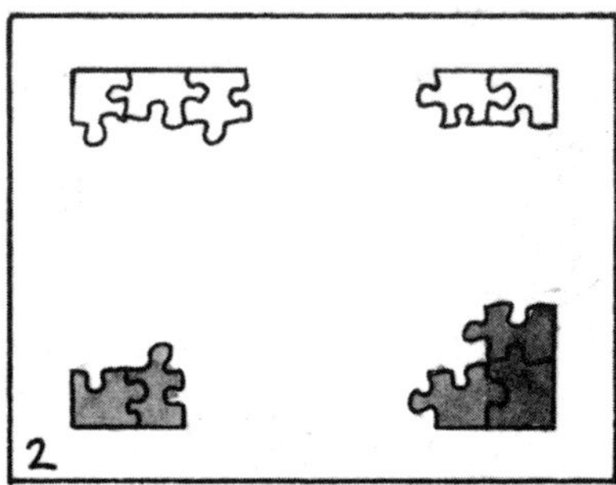
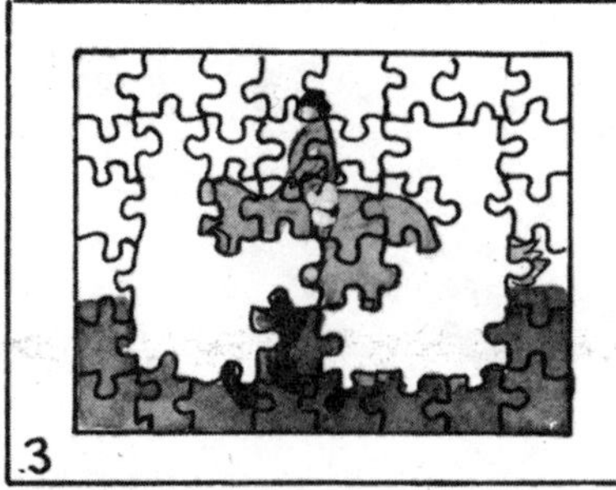
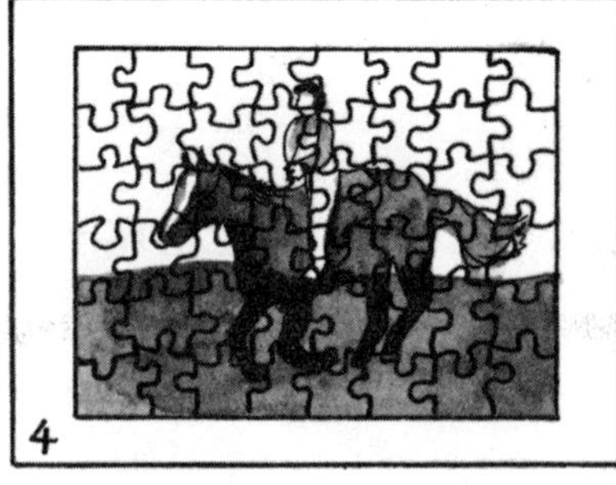

A jigsaw can seem to be magically putting itself together piece-by-piece using the animation function on a camcorder

and your camcorder must be securely mounted on a tripod. When you have tried this, you can then progress to stop-motion techniques – making a toy car seem to come to life, for instance, by moving it forward very slightly between each take.

Dutch tilts

This is a very simple trick in which the camera is twisted so that the ground or horizon appears as a diagonal, rather than a horizontal line. It is most frequently used when filming skyscrapers – making them seem a touch more dynamic. Cars moving across the frame can be shot in this way too – emphasising their speed and motion.

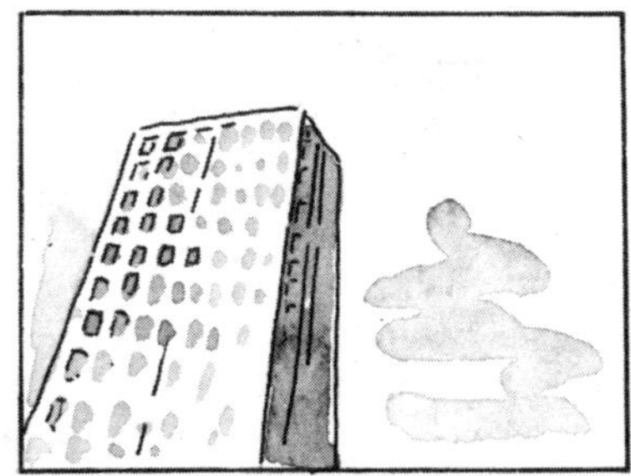
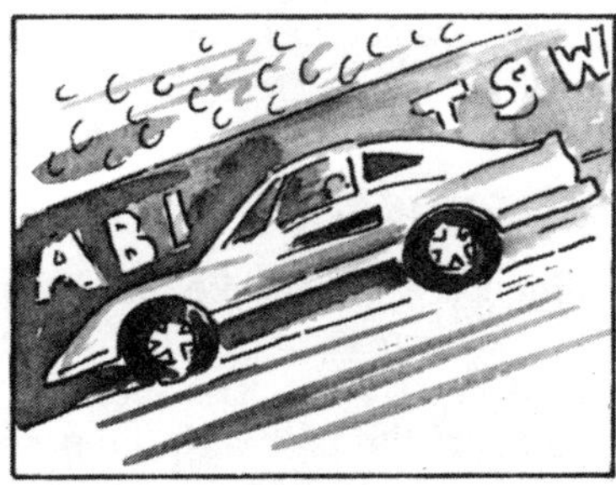

Dutch tilts are particularly suited for shots of skyscrapers and cars

Video tripods do not usually allow you to move the camera in this plane, so in order to make things on the slant you have to make two legs shorter than the third.

As with all such tricks, it is an effect that should not be over-used. Sloping horizons are disconcerting for the viewer and should only be added with purpose. See colour photograph in centre section.

Shoot yourself

One of the problems with using your camcorder to make a record of your family growing up is that there is always one person who rarely makes an appearance in front of the camera – and that is you! True, you can try to persuade and teach other family members to use the camera, but you may still find yourself being left out more often than not.

There is a simple trick for getting yourself in front of the camera without too much effort and without the need for special features such as a selftimer. The answer is to hold the camera at arm's length, with the lens pointing towards you. If you set the zoom to wideangle, you will find you can get a very presentable head and shoulders shot of yourself. What is more, it is almost impossible to tell on video that you were holding the camera.

It may take a bit of practice to be confident of getting yourself properly framed, but after that you will be able to add shots of yourself whenever you feel you are being left out of the videoed proceedings. You could use this technique for reaction shots of yourself smiling at what the kids are getting up to. Or you could introduce a new location with a short commentary to camera.

Disappearing act

Although it might take magicians years of training to make their assistants disappear on stage, for the video user this is one of the simplest tricks in the book. The secret is that you are cutting two different shots together: the first with the person in place, and the second with them out of shot. Run the two together and the person instantaneously disappears – as if by magic.

The only thing you have to watch with this trick is that nothing else moves in the frame between the two shots and this means keeping the camera perfectly still with the aid of a solid tripod. Of course, the technique can be used to make things appear as well. Using a succession of such shots, for instance, you could make a sequence showing a set of paintings mysteriously attaching themselves to a blank wall one by one.

An even more impressive form of this trick is open to those with a digital mix facility on their camcorder. With this facility, the first shot gradually fades into the second. The effect on screen is similar to something being 'beamed up' by the Starship Enterprise – an ideal trick for Star Trek fans!

Day for night

The day-for-night technique allows you to shoot realistic-looking night-time shots in the middle of the day. This can be useful as night footage can often look grainy and colourless. To perform the trick you need to do three things:

- Turn your camcorder's manual white balance control to its tungsten bulb setting. This will give your picture an overall blue tint – a colour which we all associate with night. If you have no white balance control, or need a stronger blue, you can get the same effect by putting a dark blue filter on your lens.
- Shoot into the sunlight. This will provide the high contrast between light and dark that is normal in moonlight.
- Underexpose the picture slightly. This will make the blue colouration darker and help hide detail that you would not normally see at night.

With a little care, you can even include the sun in your shot and it will appear like a full moon in your video. See colour photograph in centre section.

Split-screen

How do you record two things going on at once so that they can both be seen at the same time on screen? Without using expensive electronic gadgets, the answer is by using a simple mirror. The classic example of where this effect comes into its own is when you want to video two people having a telephone conversation with each other. If you are making a drama, it does not matter if the two people are actually in the same room.

By placing one of your subjects in front of the camera, and the other at 90° to the camera, you can get them both in shot by placing a mirror just in front of the lens. By angling this mirror at 45° and fiddling around with its exact position, you will be able to get one of your subjects filling the left hand side of the frame, and the second on the right hand side – with the straight edge of the mirror forming the border between the two.

With a little ingenuity, it is possible to adapt this technique for performing wipes between two scenes. By pulling the mirror across the field of vision, the picture on screen will appear to wipe from the direct

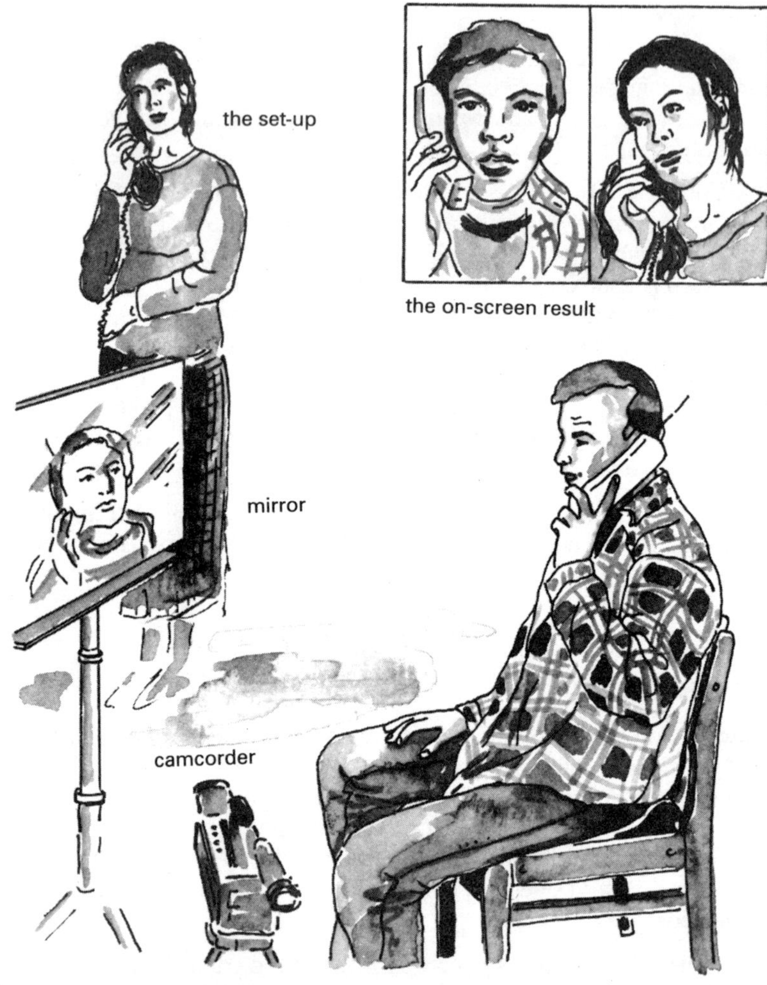

If you want to show two people having a conversation on the phone, so they appear on the screen at the same time, you have to cheat by having both actors in the same room and using a carefully positioned mirror

view to the reflected view. By changing the position of the mirror, its angle, and its direction of movement you can execute different wipe patterns.

Vignettes

A vignette or matte is a simple device for masking off part of the picture. The effect is used for simulating, say, looking through a pair of binoculars or looking through a keyhole. These vignettes are made by cutting out the required shape from a piece of black cardboard, and holding it just in front of the lens while shooting. In order to get the sharp, visible outline of the shaped hole you will need to shoot in bright light to ensure that the iris opening is small enough so that you have enough depth of field. Increasing the distance between the mask and the lens will also help make the outline look sharper. Other shapes you might want to try include a heart (for loving couples) or an oval (to simulate a Victorian photograph)

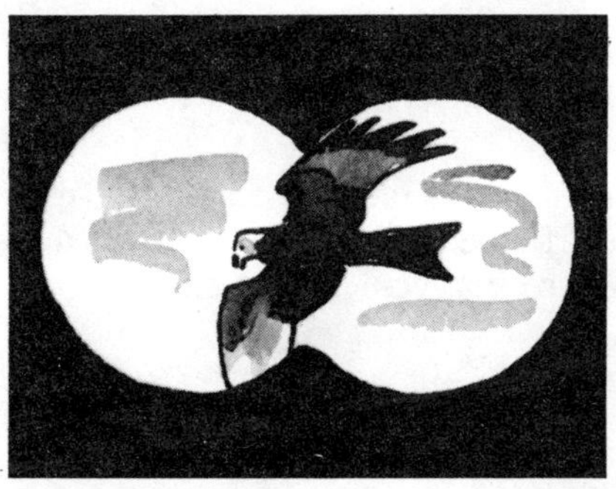

A cutout piece of black cardboard in front of the lens can make it look as if you are watching your subject through a pair of binoculars

Larger than life

As television is a two-dimensional medium, it is quite easy to fool the viewer that things are farther away from, or nearer to, the camera than they think. One of the best ways of abusing this assumption is by making toys or miniature models look life-size.

If you place a toy car, for instance, near to the camera and crop in tightly it can be made to look like a real car. The key is not to leave anything else in the frame that gives the correct idea of scale. A blade of grass, for instance, near the car will give the game away. But objects in the background, as long as they are a reasonable distance away, can help to strengthen the illusion. For instance, if a person stood far enough behind the toy car they would look small enough in the frame to be in scale with the car.

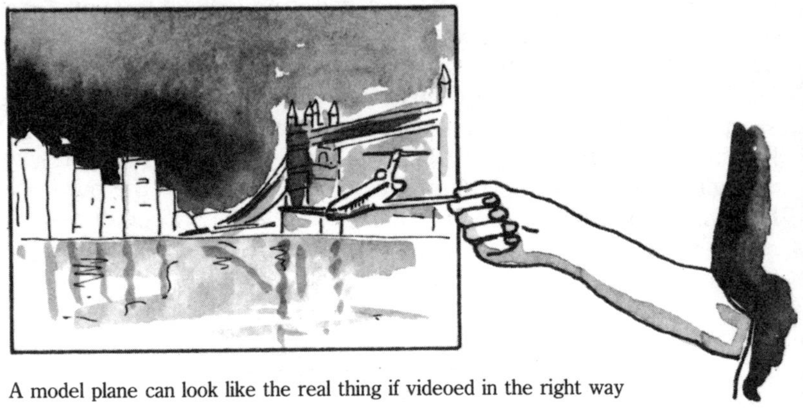

A model plane can look like the real thing if videoed in the right way

As this technique depends on your ability to pay attention to minute detail, it is best performed with a reasonably-sized television connected to the camcorder, so that you can see exactly what you are shooting.

15
EDITING

Although it is possible to make a finished video adding shot after shot with your camcorder, nearly every tape you shoot will benefit from at least some editing. For most professionals it is at this post-production stage that the real creativity begins.

In its humblest form, editing allows you to tighten up your video – removing the bad shots or the dull sequences. At its most artistic, editing allows you to manipulate events in hundreds of possible ways so that the finished video looks almost unrecognisable from the raw footage you originally shot. In Chapters 9 and 11, we looked at the techniques for putting a video together as you shoot it – 'in-camera editing' as it is sometimes known. These same rules are used when editing your video in post-production – except here you can be ruthless about what you keep and what you do not keep. You can also change the order of shots to help tell the story better.

The mechanics of editing

Unlike film, video is not cut and spliced together. The video tracks that hold the information that make each frame are so narrow, and are written on the tape at such an angle, that such a physical break would lead to a noticeable disturbance on screen, should the tape be replayed. Even more importantly, the uneven joins that would result could damage the delicate recording heads inside your equipment.

Video editing is a copying process. You play the original tape on one player (usually the camcorder) and record it via a set of leads to another tape deck (normally a VCR). You select which bits are actually recorded by stopping and starting the second machine as required.

The editing, or post-production, process goes further than just rearranging the shots. It also allows you to manipulate the soundtrack so that you can add music, commentary or sound effects to your re-recording. At this stage, you can also add other effects, such as titles, fades or mixes, to give your video a professional polish.

There is one main disadvantage to editing video tape that is not present when handling film. As film is physically cut you can add new shots to your edited film wherever you choose – you add a new bit and the rest just gets shunted along. This is not possible with video. Anything you add either has to be placed at the end of whatever you have already or it has to replace something that is already there. Careful planning is necessary, therefore, so that you don't find yourself in the situation where you want, say, to add a new opening shot to your edited movie, but have no space to put it.

Simple copying

Whether you want to get involved with editing or not, you will undoubtedly want to make copies of your tapes. Most commonly you will want to transfer what you have shot from your camcorder format tapes to a VHS cassette so that it can be watched easily by friends or relatives who do not have a suitable camcorder at their disposal. This copying process is, in fact, the basis for all editing.

For this copying process you will need your camcorder to play the tape you want to copy, your VCR to record the output from your camcorder, and a television so that you can monitor what is going on.

The first problem you will come across is wiring your camcorder to your VCR. It will be a matter of luck if you have the leads that you require. Manufacturers have little regard for standardising sockets to make this process easy. As we saw in an earlier chapter, camcorders can have one of half a dozen different combinations of AV out sockets.

Your camcorder will come with an AV lead that will have a plug (or plugs) to fit this socket (or sockets) at one end, and will most probably have two phono plugs at the other (three phono plugs if you have a stereo camcorder).

Your VCR, unless it is very old, is likely to have a SCART AV input to the rear. To connect your camcorder you will therefore need a lead that has a SCART in plug at one end, and the plug/s to fit you camcorder's AV out socket/s at the other. Alternatively you could buy a phono-to-SCART adaptor (fortunately these are now being included as a standard accessory with more and more camcorders). Other VCRs have phono inputs, which again will simplify the connection process.

If you have a high-band camcorder, you will find that you will not be able to use the S-video lead with a normal VHS VCR – as this is a low-band unit. Instead, you will have to use standard AV leads.

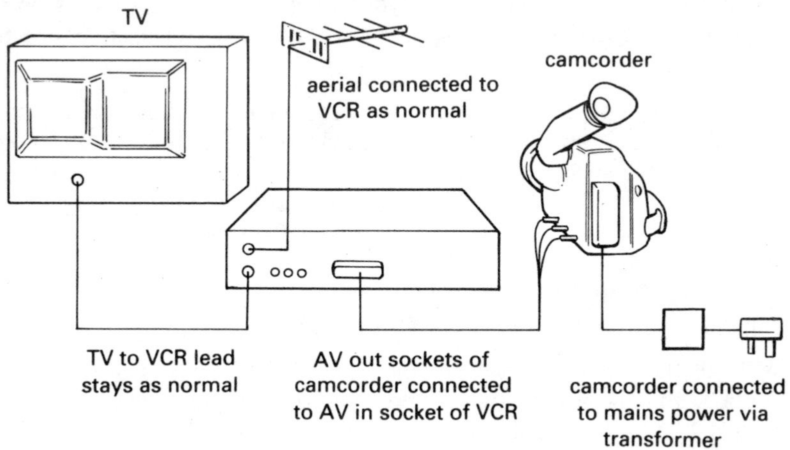

The basic set-up for copying camcorder tapes to VHS, or for simple assemble editing

Although some VCRs will automatically sense that their AV in sockets are being used, most will need to be told that you want to record from an external source – rather than from one of the preprogrammed TV channels on the VCR's tuner. Most VCRs let you do this via a switch on the VCR itself or on the remote control which is marked AV/TV. Switch this to AV (if your VCR has more than one input you will have to select the correct one from AV1/AV2 etc.) Alternatively, the AV button might be labelled 'Ext' or 'Aux' or might be found as channel '0' on the VCR.

Whatever wiring is needed, it is always preferable to connect your camcorder to your VCR directly, rather than using the RF converter. Using the RF converter gives a vastly inferior picture which will only exacerbate the deterioration in quality that you get with the copying process. With older VCRs with no AV in sockets, you may be forced down this route. To do this you connect the RF unit to the camcorder in the usual way and plug the other end into the aerial in socket of the VCR. The complicated part of the process is tuning the RF adaptor and VCR so that one can talk to the other without interference from TV channels. First tune a new channel into your VCR (you will need the VCR's instruction booklet for this). But rather than trying to find a TV station, find a channel where there is no picture at all and set this. Select this new blank channel on the VCR with your TV switched on. Then tune the RF adaptor with a miniature screwdriver with the camcorder in record pause. When the RF unit is tuned in you will see the picture from the camcorder on your TV screen. You are then ready to go – but remember, only use this route if you cannot make a direct AV connection.

For the copying process, your TV is connected to the VCR in the normal way – via an aerial lead from the aerial out socket of your VCR to the aerial in socket on your TV. In this way you will be able to see either what the camcorder is playing or what the VCR is playing – depending what you are doing at the time. When the VCR is in pause, record or stop modes, the TV will show what is coming from the camcorder. If the VCR is in play (to review what you have recorded) you will see what is on the tape in the VCR.

With all these connections completed (the first time is always the worst!), you are now ready to make your copy. If your camcorder has an edit button, switch this to *on* as this will give a cleaner copy (see Chapter 4). Put a good quality tape into your VCR and set it to record pause. Press play on your camcorder and release the pause button on the VCR. The copying process has begun.

It is now an easy step forward to making your first edits – because all you have to do to cut something on your copy that was on the original is to press the pause button on your VCR – releasing it again when the next interesting sequence comes on screen. To do this effectively, you should watch the tape before you start so you can jot down the counter numbers where you want to cut sections out. You can then be prepared to press the pause button on cue.

Simple assemble editing

Assemble editing is the technical term for selecting sequences from a tape and copying them in turn in the order you want them to appear. Although the process outlined above performs this process, there are a number of refinements that should be made to your procedure to ensure consistent results. With due care, it is perfectly possible to make a professional-looking movie with just a camcorder and a VCR – without any other editing equipment. All an edit controller does, as we shall see later, is take away some of the drudgery and some of the inaccuracies that can creep in with this manual method.

The first improvement to your editing method should be to cue up each sequence you want to copy separately – rather than letting the camcorder roll through continuously, hitting the pause button of the VCR on the hoof. Use the picture search facilities on your camcorder to find each shot you want to record. Wind back to about ten seconds before this shot – this will allow the camcorder enough time to get up to speed, to give the best possible picture. Then use the picture search facilities on your VCR to find where your recorded shot ended and put the VCR into record pause at this point. Press play on the camcorder. After the ten second pre-roll, release the pause button on the VCR.

After practising with this technique for a while, you will realise that you are losing a second or two of material from your edited version – a second or so at the beginning of the sequence, and a fraction of a second at the end. This is because of the backspacing system used on the VCR – before it starts recording it rewinds the tape slightly to ensure that you get a clean edit (with no gaps of blank tape which would show as noise on screen). It also uses this time to synchronise the incoming signal with what is already on the tape – so that the video tracks are correctly aligned at the edit point. If you press play on the camcorder and release the pause on the VCR simultaneously, you will lose information from the end of your last shot – and some from the beginning of your new one. You must allow for this by:

- always recording slightly more than you want when each shot is transferred.
- releasing the pause on the VCR about a second or so before the camcorder reaches the point where you want the edit to start from.

The exact timings will depend on the characteristics of your VCR and camcorder and you will have to learn these with practice.

Another tip for improving the look of your edited videos is to start and end them with a section of black tape. Unrecorded tape when replayed gives a noisy picture and makes your videos look unprofessional. If the picture is completely black, however, it looks less distracting on screen. You should allow about 20 seconds of black tape at the beginning and end of your video. To make this black leader you simply record some footage with your camcorder's lens cap on! You can record this directly on the VCR by putting the camcorder into record pause (with lens cap on) and putting the VCR from record pause into record. You will also be recording a soundtrack – so make this recording in a quiet room. Alternatively, if your camcorder has a mic socket, you can cut out the sound completely by putting an unconnected plug in this socket.

Sound mixing

One of the simplest ways to improve a video at the editing stage is by adding a new soundtrack. A piece of music played over certain sequences of your video, for instance, will help the shots run into each other much more easily as there will be no jumps in the soundtrack. What is more, you are able to mix the volumes of any new sound you add and the existing sound throughout your edited copy – so that you can cut down the background noise when it is not wanted, for instance.

Sound mixing does mean buying a piece of extra equipment – an audio mixer. Fortunately, this is a very cheap post-production accessory and well worth the investment. Audio mixers can also be found built into many other types of editing equipment.

The most straightforward way to mix a new soundtrack is to do it while you are editing. The equipment set-up is similar to that for straight copying or assemble editing – except the sound output from the camcorder is routed through the sound mixer on its way to the VCR. The video signal is fed straight through to the VCR as before. If you are using phono plugs, the wiring is easy to work out as you have separate sockets for the audio and video signals. Most audio mixers have phono sockets for this reason. If you have other sockets on your camcorder or VCR, you will need to get the appropriate leads or adaptors to make the necessary connections.

Typically an audio mixer will allow you to feed in three different audio inputs:

- The sound from the camcorder.
- Sound from a microphone – for adding commentary.
- Sound from another sound source – such as a CD player, or tape deck.

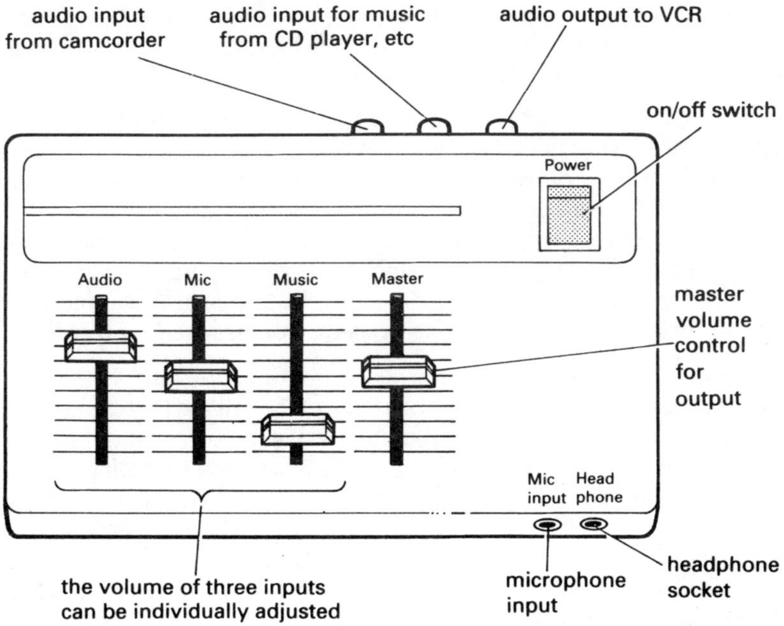

audio input
from camcorder

audio input for music
from CD player, etc

audio output to VCR

on/off switch

Power

Audio   Mic   Music   Master

master
volume
control
for
output

Mic   Head
input phone

the volume of three inputs
can be individually adjusted

microphone
input

headphone
socket

A simple audio mixer

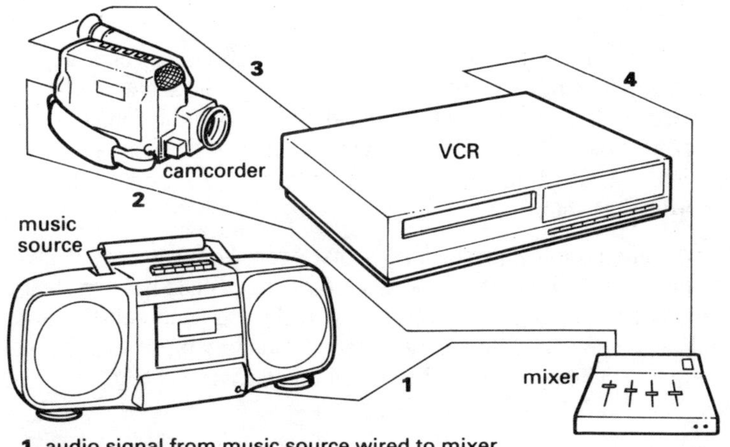

1 audio signal from music source wired to mixer
2 audio signal from camcorder wired to mixer
3 video signal from camcorder wired to video in socket on VCR
4 audio out socket of mixer wired to audio in socket of VCR

The basic set-up for mixing sound

Each of these inputs can be independently adjusted in volume to get the right mix and a further level control allows you to adjust the volume of the mixed, output signal.

Obviously, making complicated adjustments to the soundtrack while you are also trying to edit the video is going to the impossible – unless you have six pairs of hands! It also makes it difficult to add continuous sound over different edit points. For this reason, sound mixing is often done as a separate stage after editing. The process is the same, but instead of editing the camcorder footage – you are editing the edited version – putting a new soundtrack on this. Instead of using a camcorder as the replay deck in the copying process, you use a second VCR.

This process does have its drawbacks – the final tape is a third-generation recording (i.e. a copy of a copy of the original first-generation recording). What is more, if you then make copies of this tape for others to watch, you end up with a fourth-generation recording. Every generation means a drop in quality – and by the third generation the deterioration of the picture is getting noticeable. This is why those who take video seriously start off with high-band recordings, edit them using high-band VCRs, and then make only the final copies on ordinary VHS. In this way, the quality at the end is still good enough for most people.

Audio dubs

One way of minimising the number of generations of recordings you have to go through to get your finished copy is audio dubbing. As described earlier in the book, audio dub is a function found on some VHS-type camcorders, and allows you to replace the mono linear audio track without affecting the picture (or the stereo sound tracks, if the camcorder has them).

The advantage of editing to VHS is that even 8mm camcorder owners have the ability to audio dub – as long as their VCR has this feature. If the VCR has a HiFi stereo tracks as well, you can keep the original sound on these – adding the new sound (such as music) to the mono linear track. These can then be mixed on replay.

Insert edits

As on camcorders, if a VCR has audio dub, it also likely to have insert edit. This feature allows you to add new video tracks to your edited tape, without affecting the sound on the mono linear audio track. This means, for instance, you could copy a whole sequence of a band playing at a party to your VCR from your camcorder. You know the footage is boring as the music lasts a couple of minutes, but you don't want to cut it as you will end up with breaks in the music. Instead you insert shots of people dancing, drinking, and making merry, at selected intervals throughout the copied footage of the band. The result is a complete dance number which cuts quickly back and forth between band and guests – but the music remains continuous and complete. In fact, you have condensed the goings-on of a party that lasted several hours into just two minutes of fast-moving action which looks and sounds highly professional.

Edit VCRs

As we have said, you can use any VCR for editing – but you will soon discover that some are better than others. If you are going to do a lot of editing, or are about to buy a new VCR anyway, it makes sense to get one that helps, rather than hinders, you. Key features that are worth looking out for on a VCR are:

- Audio dub.
- Insert edit.
- A good range of AV inputs. Phono sockets are generally the most useful (particularly if they are mounted on the front of the VCR so that you do not have to keep reaching behind the machine).
- Jog/shuttle dial. This is the best form of tape transport control. The outer ring allows you to 'shuttle', that is to fast forward and rewind through a recording to find the shot you want. The more you turn the shuttle ring the faster the tape moves. When you stop turning the ring the picture goes into freeze frame. By turning the centre ring, the jog control, you can then advance the video by a frame at a time. It give you almost complete control over the tape allowing you to cue up the beginning of edits precisely. Unfortunately camcorders normally have a fairly primitive tape transport mechanism – making it hard to cue up the tape exactly as you want it. For this reason, professionals use a second VCR instead of the camcorder to play the tape from. A jog/shuttle dial on the VCR's remote control unit is an added bonus.
- HiFi stereo sound – particularly if you have a stereo camcorder.
- Synchro edit socket. Allows recorder to be started and stopped by another machine (see below).

- Edit control socket. Panasonic RMC or Sony Control-L remote control system – as used on camcorders (see Chapter 4). This allows a suitable edit controller to take complete command of the VCR's transport and record mechanism (see below).

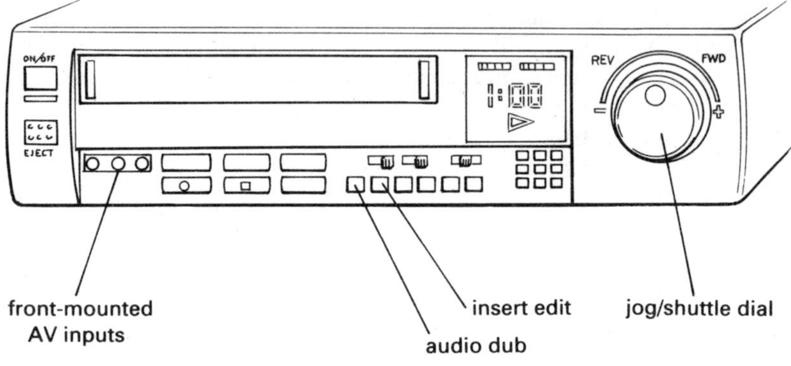

A VCR with a good range of editing functions

Synchro editing

Editing involves controlling two machines – the play deck (usually the camcorder) and the record deck. To do things manually you need to start one, and then start the other. To simplify the process slightly, manufacturers have come up with various systems that allow you to start both machines together – just by using the controls on one machine. The systems vary in detail, but generally you leave the camcorder in play pause and the VCR in record pause – then by pressing one of the pause buttons, the camcorder goes into play and the VCR into record. You then press one of the pause buttons at the end of the sequence you want to record.

The disadvantage of the system is that you are still starting and stopping the machines manually for each edit. Furthermore, for synchro edit to work, both camcorder and VCR (or both VCRs) must have the same synchro edit system – which generally means they have to be of the same make.

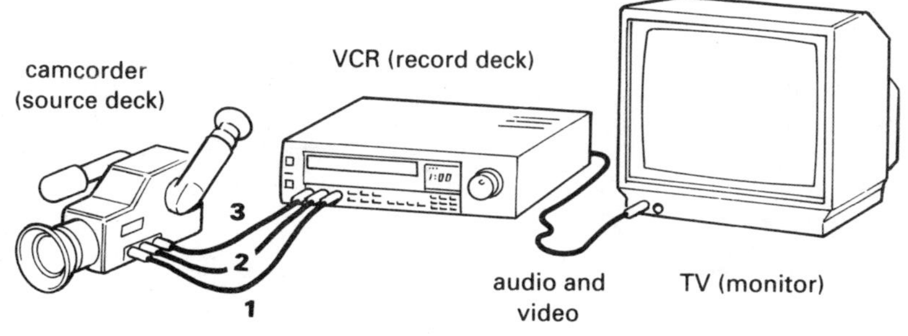

1 synchro-start remote control
2 video
3 audio

Set-up for synchro-start editing

Edit controllers

One way to take some of the strain and drudgery out of editing is to invest in an edit controller. The types available vary enormously. There are ones that only allow you to perform a single cut at a time with little more accuracy than performing the edit manually. Towards the other extreme, there are those that allow you to pre-program in hundreds of different shots you want copying, and that can be adjusted to suit the backspacing and pre-roll characteristics of both machines to ensure a high degree of accuracy.

Edit controllers allow you to perform the majority of the editing process from the one console. They control the VCR via an edit terminal or by the machine's infrared remote control system. They take charge of the camcorder via its edit terminal (either Control-L or Panasonic 5-pin RMC). Edit controllers are available for camcorders without this edit socket – but their functions and accuracy are severely limited, and you are just as well off editing manually.

What most edit controllers do is to read the camcorder's internal tape counter via the edit terminal. In this way, when you tell the editor the beginning or end of a scene, the machine can memorise these points. You can then build up your edited movie a shot at a time, and the editor remembers all the relevant stop and start times. Normally, you can

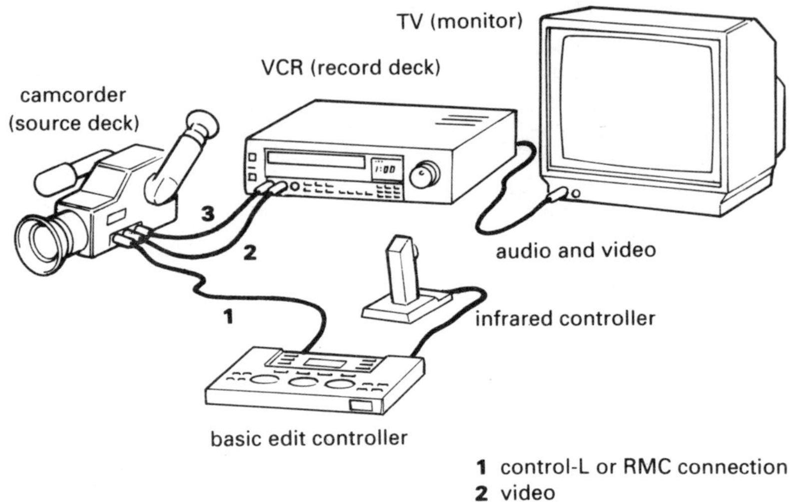

Basic editing set-up using Control-L or RMC sockets

then preview your edited movie before it is transferred to the VCR – though there will be gaps between each sequence as the controller fast forwards or rewinds the camcorder to the correct position for the next shot. You can then edit individual cuts if necessary and you can adjust the start-up pre-roll and delay times so that the VCR starts recording at precisely the right moment. Once you have made all your adjustments, you just tell the editor to make your edited copy and it gets on with it, without any more input from you. Some edit controllers will even keep complete edit lists in their memory – so you can come back and make a new version of that tape at a later date without having to start the process all over again.

Timecode editing

Using the counter readings to identify and relocate edit points is good enough for most people's purposes, but it is not entirely accurate. This is because the tape counter is just that – a tape counter. It does this by counting the number of frames from one set position to another. Unfortunately, it can occasionally miss frames – or the tape can slip – so that a counter reading for a particular point on the tape could be slightly different next time you cue up the tape. These small mistakes can be cumulative – the counter may be only five frames out to begin with, but after ten edits you are two seconds adrift. Making sure you fast wind and rewind the tape a couple of times before you start can minimise the effect of tape slippage – but you cannot get round the fact that the counter reading is not actually related to anything on the tape.

The most accurate way to edit, therefore, is with timecoded recordings. Timecoding is a way of marking each frame of a recording, so that it can be identified and located precisely by a suitable edit controller. In theory, at least, this means that each of your edits is frame accurate and there is no progressive inaccuracy, however many edits you string together. In practice, the achieved accuracy will depend much on the quality and compatibility of the camcorder, VCR and controller that you are using.

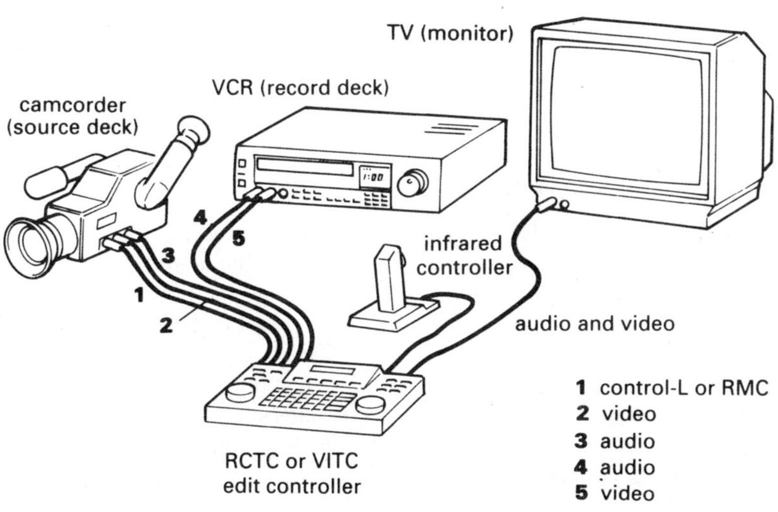

Editing using timecoded recordings

There are two timecode systems commonly found on domestic video equipment – VITC and RCTC. VITC (vertical interval time code) is found on a number of VHS-type camcorders made by Panasonic, JVC and others. The VITC generator is built into the camcorder itself, or it comes as a plug-in module. VITC has to be recorded at the same time as the picture. It cannot be added to a video afterwards – unless this is done as part of the copying process.

RCTC (rewritable consumer timecode) is found on a couple of Sony's top-of-the-range Hi8 camcorders. It has the advantage over VITC in that it can be added to 8mm and Hi8 recordings without disturbing the existing audio and video tracks and without the need for copying.

Choosing an edit controller

Picking the right editor for you will depend to a great extent on what equipment you are going to use with it. There is no point, for instance, having one that can read timecode if your camcorder does not generate timecode in the first place (unless you are going to update your camcorder in the near future). More importantly, an editor which controls a camcorder using the Control-L socket is going to be of no benefit to someone with a Panasonic camcorder.

Some edit controllers are 'dedicated' – that is, they are designed to work with a specific range of equipment (such as camcorders with an RMC 5-pin edit terminal, as found on many camcorders made by Panasonic). Others are 'universal' and work with a wide range of different camcorders and VCRs – although they will still work better in certain setups than in others.

Increasingly, edit packages are being written for computers – so that you can control your camcorder and VCR from the keyboard or mouse of your PC, Mac, Amiga, etc. The edit package takes the form of the relevant software on disk, and a minimal amount of hardware (in form of a connecting lead or command box). This route has obvious advantages for those that already have suitable computers. The computer is well suited to memorising the edit points and giving a full display showing you what you have done in the form of an edit list. These packages also work out cheaper than buying a stand-alone unit (unless you need to buy the computer as well).

16
ADDING TITLES

A title sequence is an integral part of any professional film or video. It not only provides useful information about what you are about to see, it also, perhaps more importantly, tells the audience that the video is about to start, and gives them a chance to settle down. Whatever sort of videos you shoot, there is no denying that titles will give your humblest productions a polished look. And you could even go a step further and add your own credits at the end . . .

The great thing about titles is that they can be added at any stage of making your video:

- You can shoot them before you start shooting the video proper – that way they will be in the right place, at the beginning of the tape. This will also give you the chance to prepare them before, say, you go on holiday.
- They can be inserted with the camcorder at a later stage of shooting. If you are going to do this it is worth leaving a few seconds of blank (or blacked) tape at the beginning of the tape.
- They can be added in during the final editing stage.

You can make titles in hundreds of different ways. Many camcorders come with one of two different title systems built in – either a digital superimposer or a character generator (see below). Often as effective are the home-spun titles – such as writing on a blackboard, or drawing words on a sandy beach. At the top end of the scale, computers or special units called title generators can be used to give you a wide range of typefaces, colours and effects (such as scrolling).

Character generators

Having a built-in facility that allows you to punch in letters to make your own titles might sound like an attractive feature for a camcorder. The truth is that such facilities are often very limited and will therefore look repetitive if continually used on your videos.

Generally, these character generators allow you to add two lines of type in the centre of the screen, in capital letters only, and in a choice of, perhaps, eight colours. To avoid having to have thirty different buttons on the camcorder to input the letters of the alphabet, the three or four buttons that are used have to be pushed in the correct sequence to end up with what you want. For instance, you may have to keep one button down until the letter you want appears, and then have to press another button to move the cursor to the next letter – and so on. To add the finished title to the video, you simply keep the title in the viewfinder and press record. Your title will then be superimposed on what the camcorder sees through its lens.

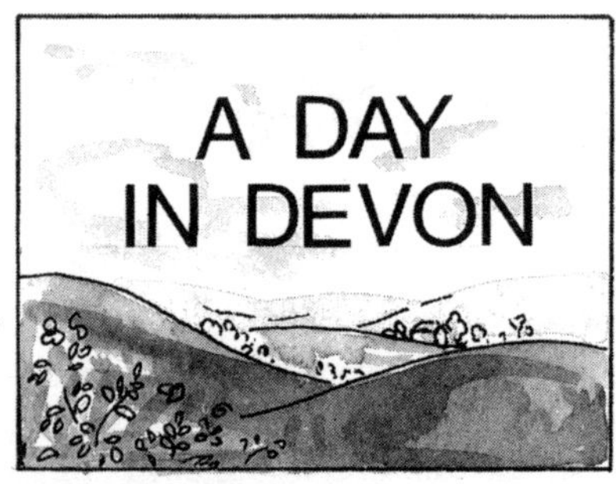

Record titles over a plain background – such as an area of sky

Although using this kind of titler is fiddly, you do have the chance to punch in the title you want well before you start shooting. The camcorder will generally remember the last title you punched in until you change it (the power is supplied by the button cell that runs the built-in clock and calendar).

Some camcorders have character generators available as optional extras. These plug into a special socket on the camcorder, and have the advantage that they have more keys – so inputting the letters is more straightforward.

To make this sort of title look effective, it should generally be shot on a plain background. An area of sky might be an ideal choice. Avoid putting titles over highly detailed backgrounds wherever possible, as they will be much harder to read. If stuck, you could always record your titles against a black background, by keeping your lens cap on.

Choosing the right colour for your lettering is made difficult because of the black and white viewfinders that are predominantly used on camcorders. The colour you have selected is only indicated by a letter in the viewfinder – such as R for red, or Y for yellow. You will have to try out these colours at home using a colour television so that you are sure you know what they all are. When selecting a colour for your titles, it is best to ensure that it contrasts well with the background colour – for example yellow lettering will stand out well against a deep blue sky, and red will look good against black.

Titles should not be kept on screen for too long. For short titles, two or three seconds is plenty of time for people to read what is going on. For longer titles, allow one second for every two or three words. An old trick for counting these lengths reasonably accurately in your head is to say to yourself 'Mississippi one, Mississippi two, Mississippi three . . .' and so on – with each 'Mississippi' equalling a second.

Digital superimposers

In some ways, a digital superimposer is not as sophisticated as a character generator. All it does is to memorise whatever is placed before the camera in a high-contrast single-colour form. It relies on you either to write the titles yourself and then input them to the camera with the superimposer, or to find ready-written titles and copy them in. However, as you are not relying on a simple typeface, or selected sizes, the superimposer does have more creative possibilities.

All you have to do to work the superimposer is to place the title you want to store in front of the camera and frame it up as you desire. You then simply press the memory button. This stores whatever is in front of the camera as a high-contrast image made up of black and white. It is kept in the store until you replace it with something else, or the button cell of you camcorder runs down. When you want to use the title you press the title button to display the stored image in the viewfinder again. You then decide which colour you would like the dark parts of this stored image to appear on screen – there are usually eight to choose from. To record the image on tape you press the record button on your camcorder – the dark parts of the stored image are recorded in the selected colour, and the light parts form a window, through which you can see whatever the camcorder is seeing.

The beauty of the system is that you can take advantage of artwork that has already been printed for another purpose. For example, you could memorise your company's logo from a sheet of headed notepaper or you could memorise the front of the order of service to make a suitable title for a wedding video. Similarly the wording from a road sign or guide book could be memorised to title a holiday video.

When you want to make up your own title, all you need to do is to write the title on a piece of white paper and then memorise this.

The important factor in getting a good title is that the original that you are copying should be of as high a contrast as possible. If you write with a blue pen on a yellow piece of paper, for instance, it is likely that some of the yellow will be stored as black, and some of the blue will be memorised as white. The result would be a messy looking title. If you are writing the title yourself, you can use white paper and a thick, black pen to avoid this problem. If you are copying existing artwork, you will have to be more careful. Increasing the lighting may help or you may have to crop in more tightly with your zoom lens. Fortunately, you only need to press the memory button once in order to re-input the title for a second time.

Other features may also be possible with your superimposer:

Reverse This changes the black parts of the memorised image to white and vice versa. In this way, your title becomes a keyhole through which the camcorder looks.

Shadow This adds a secondary version of the image below the first in a contrasting colour, which looks like a shadow. This makes the title look three-dimensional

Second title Some camcorders allow you to store a second title, so that you do not have to lose your original image in order to be able to store another. For instance, you could keep your company logo stored permanently in the camcorder.

Scroll This makes your titles move. Instead of having your titles appear from nowhere, they run in from the side or bottom of the screen.

Improvised titles

If your camcorder has no form of built-in title generator you will have to use your ingenuity to come up with interesting and appealing titles. The only limits to this are your own imagination and artistic ability. Here are just some of the ways that it can be done:

Writing. The simplest way is to write your titles on a piece of paper as neatly as you can. Other materials can be used, however. For instance, a small blackboard could be written on in chalk for the title for a school sports day. Writing in the sand could be a good way of starting off a video of a seaside holiday.

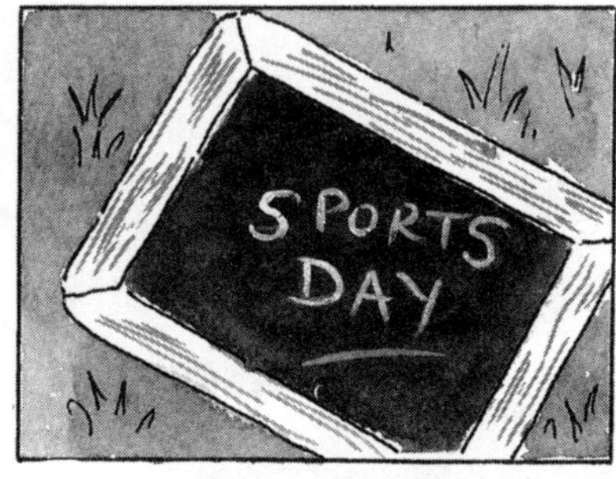

Titles on a blackboard suit school-type subjects

Rub-on lettering. For a more professional-looking title you can use rub-on lettering, such as that made by Letraset. Do not get carried away with the vast range of type styles that are available. Choose one that is bold and simple and avoid the temptation to mix lots of different styles together in the same title sequence. One style will work best. To use this lettering, you should draw a faint line with a ruler in pencil on the paper you are using so that you can line up the characters properly.

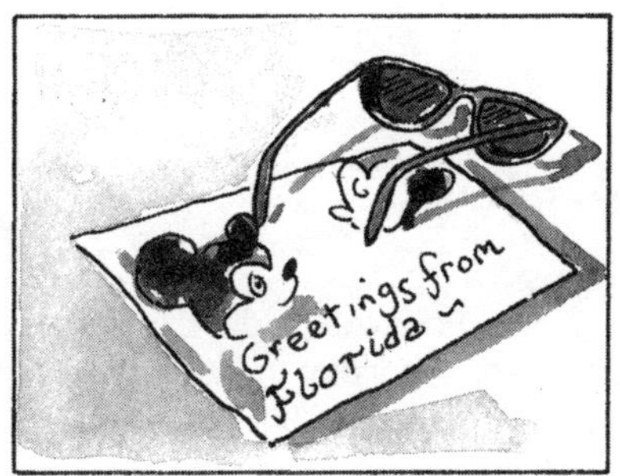

Postcards make superb ready-made titles for holiday movies. Add props, such as sunglasses, or a passport, to make the image more three-dimensional

Postcards. A straightforward way to start off any holiday video is with a still-life shot of a postcard, or a guide book which shows where you are going. A pair of sunglasses and a passport could be added to make an artistic grouping.

Signs. Another good way to start off a holiday video is with a roadsign. For instance, a three-second shot of a sign saying 'Welcome to Florida' could set the scene for your video of Disneyworld. Signs can also be used as a form of establishing shot for each day's shooting – so that you do not forget where you were when you shot that particular sequence.

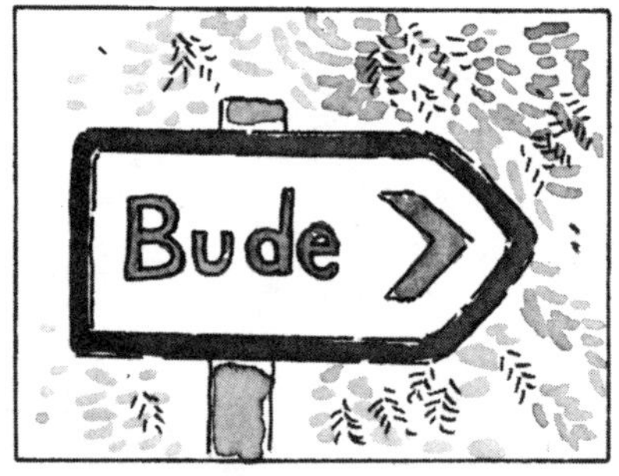

Road signs can be used to introduce a new location

Building blocks. Children's alphabet blocks are a great way of introducing a video of the kids.

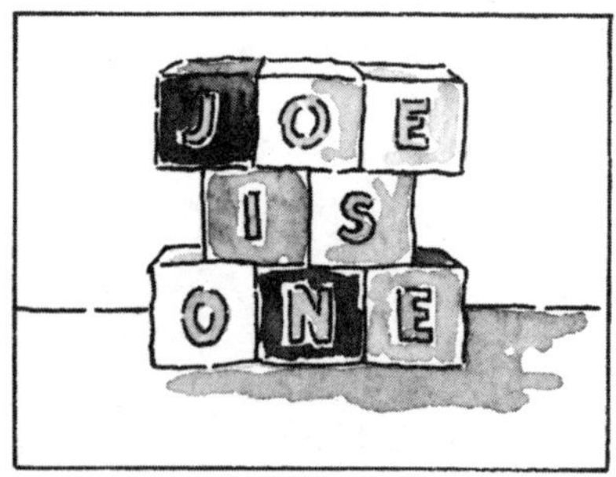

Children's alphabet blocks are well-suited for use with videos of the kids

Shells, beads, beans and things. For a title with a difference use objects to spell out letters. For instance, for a holiday video you could lay out shells in the sand so they spelt out 'Holiday '94'. For a video of your garden you could use flower petals.

Cards and invitations. Special occasions, such as weddings, can be simply introduced with a shot of the printed invitation. You could kick off Christmas and birthdays with a close-up of a suitably phrased greeting card on the mantelpiece.

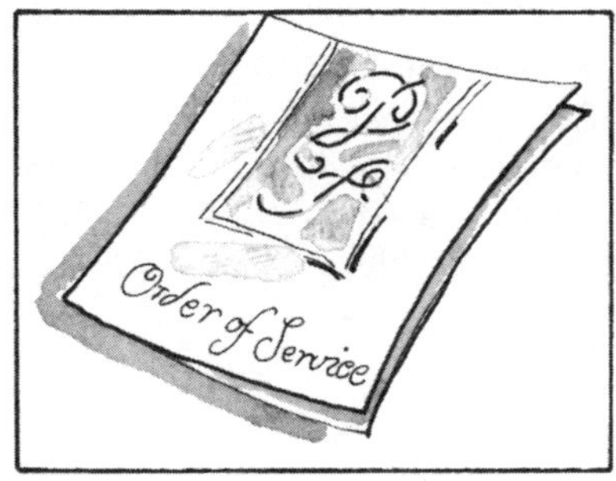

A close-up of the order of service could be used to start a video of a wedding

Stand-alone titlers

Stand-alone title generators are designed to be used while editing your videos. You plug them between your camcorder and VCR so that you can add titles to your edited copy. They vary widely in ability (and price). The simplest give only one typeface, a limited selection of colours and a limited memory. The most sophisticated give a whole library of typefaces, thousands of colours, can memorise hundreds of pages of type, and can add a whole range of special effects – such as wipes, scrolls and fades. Most have a standard typewriter keyboard for inputting characters.

Computer titling

Those familiar with modern word-processing and desktop publishing will know how easy it is to create titles using the computer. Although using this method to generate titles for your videos will be expensive for those who do not already have a computer, if you do have one, it could be an easy way of getting high-quality results.

As well as your computer you will also need two things:

- A suitable software program for compiling your titles on the computer screen. Such software is available for most types of computer – even some of the older machines.

- An encoder or genlock to convert the computer screen's picture into a normal television set picture. An encoder acts as a straight translator, so that you see what is on your computer on tape (not necessary with all types of computer). A genlock goes one step further as it also allows you to mix the type from the computer screen with your existing video as you copy it on a VCR.

At present, the greatest range of titling software and genlocks is available for the Commodore Amiga – simply because of this computer's low cost, and suitability of handling graphics. However, the range of such add-ons for other types of computer (such as IBM PC, Apple Macintosh, Archimedes, etc) is increasing quickly.

GLOSSARY

Aerial socket   Socket on television that receives high frequency waves picked up by aerial. Can be used as an input for replaying camcorder footage if an RF adaptor is used – although this degrades the picture quality.

AGC (automatic gain control)   Circuitry inside camcorder that electronically boosts the signal from the imaging chip in lowlight to get a better-exposed, though more grainy, picture.

ALC (automatic level control)   Circuitry inside camcorder that adjusts the recorded volume of the recorded sound – boosting quiet sounds and cutting down loud noises.

Angle of view   A measure of how much of a scene a lens can see from a particular position.

Animation   An effect that brings inanimate objects to life. It is done by shooting a subject in a series of slightly different shots, each lasting a very short time.

Aperture   Adjustable opening of lens that is used to control how much light enters the camera. It also has an effect over depth of field. Also known as an iris or diaphragm.

Aspect ratio   The relationship between the width and height of a picture. The aspect ratio of a normal TV set is 4:3.

Assemble edit   Editing method where shots or sequences (both sound and pictures) are copied from one tape to another in the order they are to appear on the final tape.

Audio   The sound part of a video recording (from the Latin for 'I hear').

Audio dub   The replacement of one of the soundtracks on a video recording without affecting the picture. Only possible with some camcorders and VCRs.

Audio mixer   A device for mixing two or more different sound sources together.

Autofocus   A system for automatically focusing a camcorder's lens so that the picture appears sharp.

Automatic exposure   A system that automatically adjusts settings on the camcorder to ensure the picture is neither too bright nor too dark. Iris, gain and shutter speed settings may be adjusted to make this possible.

Autotracking   System for ensuring that the video recording heads in a camcorder or VCR are correctly aligned to the existing video tracks.

AV (audiovisual or audio/video)   Equipment that carries audio and video signals.

AV output/input   Sockets that connect the audio and video signals between different pieces of equipment.

AWB (auto white balance)   Circuitry in a camcorder that adjusts the colour of a picture in any given lighting situation to look normal to the human eye.

Backlight   Light source which is behind the subject – such as when you shoot towards the sun.

BCU (big close-up)   A shot where the face of your subject fills the screen.

Blacked tape   Video tape that has been recorded on without a picture. This gives a professional looking start and finish to a video with a black screen, without picture noise. It also means that the tape has a sync signal, which can be useful when editing.

BLC (backlight compensation)   Button for use in backlight situations which increases the amount of exposure given to the picture – to help avoid silhouetted subjects when shooting against a window, for instance.

BNC   Bayonet sockets and plugs used on some equipment for sending the video signal.

Boom   A long stick or pole to which a microphone is attached so that it can be placed close to the subject without the microphone being visible.

Cardioid microphone   Microphone which records more from in front of it than behind it – with a heart-shaped directional characteristic, hence the name.

CCD (charge coupled device)   The miniature electronic chip inside the camcorder that converts light into an electrical signal. It is made up of hundreds of thousands of light-sensitive picture elements, or pixels.

Character generator   Built-in or add-on device which creates title lettering and superimposes it on the picture being recorded by the camcorder.

Chromakey   Professional trick for seamlessly inserting part of one video picture into another – used for special effects, such as making Superman fly.

Chrominance (C)   The colour part of a video signal.

Close-up (CU)   A shot that concentrates on a relatively small part of a subject. When shooting people, it refers to a shot where the head and shoulders of the subject fill the screen.

Colour balance   Alternative term for white balance.

Colour temperature   Measurement of the colour of light, often expressed in Kelvin. The human eye adjusts for colour temperature most of the time without our noticing it; on a camcorder the automatic white balance system performs the same adjustment.

Component video   A video signal where the luminance (brightness) and chrominance (colour) parts of the picture are separated. Found in high-band video formats such as Hi8, S-VHS and S-VHS-C.

Composite video   A video signal where luminance and chrominance are combined. The opposite of component video. Used on low-band video formats such as VHS, VHS-C and 8mm.

Continuity   Conformity between successive shots in lighting, direction of movement, appearance, props, etc.

Contrast   The difference in brightness between the lightest and darkest parts of a picture.

Contrast ratio   A relationship between the extremes of brightness in a picture. Camcorders can cope with a contrast ratio of about 1:30. However, some subjects are more contrast (such as a person standing against a sunlit window where the ratio might be 1:1000) so some detail is lost on the video. The human eye is capable of coping with a contrast ratio of about 1:100.

Control-L   Remote control socket found on some camcorders and VCRs (notably those made by Sony) that is used in editing set-ups for controlling tape transport via an edit controller. It can also be used to control other camcorder functions, such as zooming.

Control track   The part of a video tape that records the sync signals – these ensure that the recording heads are synchronised with existing video tracks, and ensure that tapes are played at the correct speed.

Crab   Camera movement where the camcorder operator moves sideways while shooting.

Crossing the line   A change in camera position between successive shots which makes it look as if the subject has changed their direction of movement, or the way they are looking. As this looks confusing for the viewer, it should be avoided.

CRT (cathode ray tube)   The picture tube used in most television sets.

Cut   The joining of two different shots on tape, one directly after another.

Cutaway   A shot which shows us something that was not featured in the previous shot, but is somehow relevant to it. A useful device for separating two very similar shots or for creating a visual break.

Day-for-night   Special effect in which daylight is made to look like night-time. It is achieved by setting a tungsten white balance, or by using filters, to make everything look blue, then the scene is deliberately underexposed.

dB (decibel)   A logarithmic scale for comparing different levels of power. A doubling of power level is an increase of 3dB, a thousand-fold increase in power level represents 60dB. Used in the measurement of signal-to-noise ratio, sound volume, sensitivity of microphones, etc.

DC (direct current)   Type of electrical current used to power all camcorders. It can be supplied by batteries or by a mains transformer.

Depth of field   A measure of how much of a picture is in focus – from the nearest point in the scene to the camera that looks sharp, to the furthermost point that looks sharp. Depth of field is dependent on iris setting, subject distance, and focal length of the lens.

Dew sensor   Detection device inside most camcorders that shuts down all power when dampness or humidity is detected. This prevents the tape sticking to the recording heads and causing damage.

Digital   A signal which has been processed and encoded as a series of ones and noughts. Digital technology is used in processing the picture signal on some camcorders, but as yet it is only used for recording the video signal on a few professional video cameras. A couple of domestic camcorders record sound digitally (those with PCM sound).

Digital effects   A whole range of effects, useful and not so useful, are possible using digital technology. Examples include: wipes, mixes, slow shutter speeds, electronic zooms, time base correctors, picture-in-picture and image stabilisers.

Digital signal processing (DSP)   Used on an increasing number of camcorders to offer certain special effects and to cut down on the overall number of components needed.

Digital superimposer   Facility on a camcorder adding graphics and titles to a video. The camera memorises titles placed in front of the lens and these can be called back on screen and superimposed over the video picture when required.

DIN socket   Large family of circular, multi-pin sockets with between three and 14 connections. Once common on audio and video equipment, now found as AV connections on German editing equipment. Mini-DIN 8-pin sockets are also used as AV out connections on JVC camcorders.

Dioptric adjustment   Adjustment on camcorder viewfinders that allows you to correct the distance between lens and viewfinder screen to suit the user's eyesight.

Directional microphone   Microphone that is more sensitive to sounds from in front of it, than to sounds from the left, right or rear.

Discharger   Accessory that ensures that a rechargeable battery's voltage is reduced to an ideal level before charging commences. This helps prolong the life and optimum performance of the battery.

Dissolve   Special effect in which one shot slowly appears as another disappears. Performed using a vision mixer.

Dolly   Set of wheels that fix to the bottom of a tripod that allow it to move smoothly while shooting.

Dropout   A dropout is a form of noise appearing momentarily as white or black specks or lines in the picture during playback. They are caused by imperfections in the tape, or by dust on the tape's surface – causing a momentary signal loss in that area.

Dubbing   The technical term for recording a video (or soundtrack alone) from one tape to another.

Edit   Editing is the cornerstone of professional video-making. It allows you to improve on the original camcorder footage by cutting out unwanted material, re-ordering sequences or shots, and changing the soundtrack. It is done by copying material from one tape to another electronically – from a camcorder (or video player) to a video recorder.

Edit controller   A device from which you can control both the source deck and recording deck during editing.

Edit switch   Switch found on some camcorders that switches off the picture enhancement circuitry when using the camcorder as a source deck for editing. This is necessary as most VCRs make this enhancement when replaying tapes anyway and a double enhancement would not be desirable.

Effective focal length   The equivalent focal length on a 35mm SLR stills camera which gives the same angle of view and magnification as the focal length of the camcorder. Useful as direct comparisons between focal lengths on different camcorders is made difficult as the focal length is dependent on the size of the CCD chip (four different sizes are commonly found on today's camcorders).

Electret microphone   A design of microphone found on most camcorders.

EP (extended play)   Facility found on American and Japanese camcorders and VCRs using the NTSC television system, which allows you to play and record tapes at a third of the normal speed – therefore allowing you to triple the normal maximum recording time of a tape. Only found on VHS-family format equipment.

EVF (electronic viewfinder)   TV screen found on most camcorders used to watch what you are shooting and for reviewing what you have just shot.

External microphone socket   Socket found on some camcorders, necessary for using a different microphone to the one built into the camcorder. Plugging a microphone into this socket turns off the built-in microphone.

Eyeline   The direction in which a subject's eyes are looking on screen.

Fade   To make the picture, sound or both gradually disappear or appear. A fade function is found on most camcorders and turns the picture to or from plain black or white.

Field   Half of one television picture – lasting 1/50th second (1/60sec on the American NTSC television system). A field contains alternate lines of the picture – the next field contains the missing lines from the first. The two fields make up a frame.

Fluorescent   The technical description for strip lighting.

Flying erase head   Before you can re-record on a tape, the old recording has to be erased. A flying erase head is positioned on the rotating head drum of the camcorder or VCR, and can erase the section of tape needed precisely, so there is no picture disturbance before or after the new recording. Fixed heads are not as effective and cause such picture distortion at each end of the new recording.

FM audio   A soundtrack recorded beneath the video tracks in diagonal stripes. Used on VHS-family stereo camcorders, and on all 8mm and Hi8 models. It gives a HiFi quality sound recording with a wide frequency response. FM stands for frequency modulation.

Focal length   Measure (in millimetres) of the magnification and angle of view of a given lens setting. Focal length is also dependent on CCD size, so the focal length of different camcorders can only be directly compared if they have the same size CCD.

Frame   A single video picture – lasting 1/25sec (1/30sec on the American NTSC television system). Made up of two fields.

Frequency   Used to describe sounds or electrical signals, and measured in cycles per second or Hertz (Hz) (1Hz = 1 cycle per second).

Frequency response   The range of frequencies of sound that an audio system is capable of recording.

Gain   Amplification of an electrical signal. Many camcorders have a gain up mode which is used in low light to boost the picture signal. This makes the picture easier to see, but the amplification process also increases the noise, so the picture is more grainy than usual.

Generation   Each successive copy of a tape is called a generation. The original camcorder tape is the first-generation, the edited tape of this on VHS is second-generation, and a copy made of this is the third-generation. The picture and sound quality will deteriorate between each successive generation.

Genlock   A device used to translate pictures or graphics from a computer screen to video, or for mixing a computer image with a live video image.

Helical scan   Recording system used on all camcorders and VCRs. It uses a rotating head to write the picture signal in diagonal stripes along the moving tape. Also used for HiFi audio tracks on some camcorders.

Hertz (Hz)   Unit used for measuring frequency.

Hi8   High-band version of the 8mm camcorder format which records luminance and chrominance separately on special tape to give a high-quality image with up to 400 lines of resolution.

HiFi (high fidelity)   Any high-quality sound system (mono or stereo) that has a frequency response of 20–20,000Hz or better.

High band   General term for all video systems with a superior picture recording capability – i.e. S-VHS, S-VHS-C and Hi8.

High speed shutter   Facility that allows the shutter speed on a camcorder to be set at a faster shutter speed than the usual 1/50sec. However, the number of frames shot per second remains the same – so footage shot at higher shutter speeds can look jerky when played back at normal speed. The facility is designed for giving sharp freeze frames or slow-motion footage.

HQ (High Quality)   Picture enhancement circuitry found on most camcorders and VCRs used for sharpening the image on playback.

Hypercardioid   Type of microphone which picks up sounds predominantly from a 90 degree arc in front of it.

Infrared autofocus   Autofocus system that sends out beams of infrared light to measure subject distance.

Insert edit   Facility found on some VHS-family camcorders and VCRs that allows you to re-record over the picture without affecting the mono linear soundtrack.

IR (infrared)   Invisible wavelengths of light, used on video equipment for sending signals over short distances – such as in autofocus systems, or in remote control units.

Iris   Adjustable opening in the lens of a camcorder used to regulate how much light reaches the CCD image sensor.

Jitter   Picture fault caused by the tape not running smoothly over the recording heads. Shows up as a shaky picture on playback.

Jog/shuttle   Twin-ring control found on top-quality video recorders – used for finding a particular frame on a tape quickly. The shuttle (outer) ring will fast forward or rewind the tape at varying speeds, depending on how far the ring is turned (and in which direction). The central jog ring is for fine adjustment allowing you to go through the tape a frame at a time. A very useful facility when editing.

Jump cut   A cut between two shots where the subject appears to jump, inexplicably, on screen. Generally to be avoided at all costs.

Kelvin   Unit used for measuring colour temperature.

Lanc   A type of edit socket. Also known as Control-L.

LCD (liquid crystal display)   A type of information display panel. Colour LCD panels are now being used as an alternative to the traditional black-and-white CRT viewfinder.

Linear audio   Sound track that is recorded by a stationary head along the side of the tape. Used for the mono audio track on all VHS-family recorders and camcorders. HiFi and stereo soundtracks, however, are recorded using the helical scan system.

Line input   Sockets found on very few camcorders that allow you to record pictures and sound directly from another video player.

Lithium ion   Type of rechargeable battery – lighter than nicads, less prone to memory effect, easier to monitor its state of charge, and more expensive. Pioneered by Sony.

Long shot (LS)   Shot in which a human figure occupies about half to two-thirds of the screen height.

Low-band   General term for all video formats which are not high-band – i.e. VHS, VHS-C and 8mm.

LP (long play)   Feature found on many camcorders and VCRs that halves the normal tape speed during playback or recording. It doubles the normal maximum recording time of a tape – although the quality will suffer slightly.

Luminance (Y)   The brightness component of the video signal (the black and white part of the signal).

Lux   Unit of measure for light intensity.

Macro   Facility on most camcorders for shooting very close to a subject – so the subject appears larger than life on the TV screen.

Manual iris   Facility for opening and closing the iris on a camcorder manually to change exposure – found only on a few camcorders. All camcorders have an automatic iris facility.

Manual white balance   Facility for altering the white balance on a camcorder manually – either by choosing one of a number of presets, or by taking a reading from a white surface in front of the lens. All camcorders have an auto white balance system – only a few have one of these manual controls.

MCU (medium close-up)   Shot in which a human figure appears on screen from the chest upwards to just above the head.

Memory effect   Problem associated with nicad batteries, caused by repeatedly recharging the battery before it is discharged to its optimum level. It manifests itself by batteries lasting for a shorter and shorter time between recharges. Avoided by only recharging batteries once they will no longer power the camcorder. Some battery chargers have a built-in discharger (or refresh facility) to avoid this problem. Dischargers can also be bought as accessories. Slow, overnight chargers can also be bought to help eliminate the problem.

Metal evaporated (ME)   High quality Hi8 format tape.

Metal oxide   Type of tape used by VHS-family camcorders and VCRs.

Metal particle (MP)   Type of tape used by 8mm and Hi8 camcorders.

Mixed lighting   Situation in which there are two or more light sources of differing colour temperatures.

Mixer   See audio mixer and vision mixer

MLS (medium long shot)   A shot in which a human figure practically fills the full screen height, from just below the feet to just above the head.

Monitor   A television screen used for watching video footage (therefore one that does not necessarily have a TV tuner).

Monopod   An extending leg that attaches to a camcorder to give extra support when shooting – a one-legged tripod!

MS (medium shot or mid-shot)   Shot in which a human figure appears on screen from just below the waist upwards to just above the head.

Neutral density filter   Grey-coloured filter that cuts down the amount of light entering the lens. Useful in very bright conditions, or when you want to force the iris to open to restrict depth of field.

Nicad (nickel cadmium)   Type of rechargeable battery most commonly used on camcorders.

Nickel Hydride   Type of rechargeable battery. More efficient than nicad types, and less prone to memory effect and more expensive. Pioneered by Panasonic.

Noise   Unwanted interference in audio or video signals.

NTSC (National Television Standards Committee)   Colour television standard used in Japan, USA and some other countries. Uses 525 lines and 60Hz.

Omnidirectional mic   Type of microphone that picks up sound equally well from all directions.

Overexposed   Picture in which too much light has passed through the lens, giving a white, washed-out look.

PAL (Phase Alternating Line)   Colour television standard used in the

UK, most western European countries, Australasia, much of Asia, and some other countries. Uses 625 lines at 50Hz.

Palmcorder   Panasonic trade name for a range of palm-sized camcorders.

Pan   To move the camera in a horizontal arc while shooting.

PCM (pulse code modulation)   A digital HiFi stereo sound recording system found on a few Hi8 camcorders. As it is recorded on a separate section of tape, using the helical scan system, it can be audio dubbed.

Peritel   Another name for the Scart plug/socket.

Phono plug   Widely used for audio and video connections. Separate phono plugs are used for composite video, left channel audio, right channel audio, and mono audio.

Pixel (picture element)   A single light-sensitive cell in a CCD image sensor. Each CCD has between 300–500,000 pixels.

Polariser   Filter that only lets in light vibrating in one plane. As such it can be used to deepen the colour of part of a picture, such as the sky. It can also be used to eliminate reflections from non-metallic surfaces such as water or glass. It must be rotated in front of the lens until you get the desired effect in the picture.

Post-production   General term for the editing process. Everything that is done to a video after shooting – including cutting, sound mixing, titling, picture enhancement, processor effects and copying.

POV (point of view)   Camera angle that shows you exactly what the subject in the last shot can see – from their point of view.

Pre-roll   Time that it takes for a camcorder or VCR to get from pause to play or record.

RCA socket/plug   Same as phono plug/socket.

RCTC (rewritable consumer timecode)   Sony's timecode system used on some of its Hi8 camcorders.

Reaction shot   Shot that shows someone's reaction to something that has happened in the previous shot – such as a shot of someone smiling.

RF-adaptor   A unit that converts the composite video and audio signals from a camcorder into radio frequencies which can be input into the aerial socket of a TV or VCR.

RGB (red-green-blue)   High-quality picture signal with separate red, green and blue components. Used mainly for colour computer screens.

RGB converter   Device that transforms S-Video signals to RGB signals for connecting high-band camcorders to TVs that do not have S-Video sockets, but which do have Scart sockets capable of accepting RGB signals.

Scart   A 21-pin AV connector commonly used on televisions and VCRs. Not all Scart sockets are the same – some are wired for S-Video, some are wired for RGB, some are wired for ordinary composite video, some are wired for stereo sound. To add to the confusion, some are AV in sockets, some are out sockets, others can be both. Special care, therefore, must be taken to use a Scart plug that is correctly wired for the socket you are using, and for what you want to do.

Scroll   To move titles up, down, or sideways, across the screen – so that they appear and disappear at the edges.

SECAM (Séquential Couleur à Mémoire)   Colour television standard used in France, some parts of Eastern Europe, some parts of the Middle East, and some other countries. Uses 625 lines and 50Hz.

SEG (special effects generator)   General name for processors that produce special effects such as wipes, mixes, chromakey and solarisation.

Sequence   A series of linked shots.

SFX   Abbreviation for special effects.

Shot   The basic unit of a video, which lasts from when you press the record button on the camcorder until you press it again to put the camcorder into pause.

Shotgun microphone   Unidirectional microphone of the hypercardioid or supercardioid design.

Signal-to-noise ratio (SNR)   A measure of the proportion of noise in video and audio signals. Measured in decibels (dB). The higher the signal-to-noise ratio, the higher quality the signal or recording.

SP (standard play)   The normal playback and recording speed on camcorders and VCRs – as opposed to LP or EP.

Storyboard   A series of drawings with captions, used to plan out a video before it is shot.

Supercardioid   Unidirectional microphone which accepts sound, primarily, in a 120° arc.

S-VHS (Super VHS)   High-band version of the VHS tape format which records luminance and chrominance separately on special tape to give a high-quality image with up to 400 lines of resolution.

S-VHS-C (Compact Super VHS)   Compact version of the S-VHS format – using the same width tape in smaller cassettes. The high-band version of VHS-C.

S-Video   Type of socket and plug used on high-band equipment as a video input or output. It is a 4-pin mini-DIN design, and keeps the luminance and chrominance signals separate. Usually used in conjunction with one or two phono sockets (two if stereo) to connect the audio signal.

Synch signal   For each frame that is recorded on the video tape there is a synch (synchronisation) signal – a simple pulse that tells the camcorder where each track is when playing back the tape. On VHS-family tapes the synch signals are found on a separate linear track. On 8mm and Hi8 tapes the synch signals are recorded as part of the diagonal video track.

Synchro edit   A simple editing system in which a camcorder and recorder can be linked together via a special lead so that both machines can be controlled from one button. So when you press the play button on the camcorder, the VCR automatically goes from record pause to record. It will only work for a single edit at a time. The synchro edit system on one piece of video equipment is not automatically compatible with another. Variations on the same theme can be found called remote pause, duet edit, and master edit control.

Tally light   Light on a camcorder that lights up when recording.
TBC (time base corrector)   System that generates new sync pulses to replace weak ones. Typically it straightens out the wobbly vertical lines that can appear on old recordings. TBCs are built into a select few camcorders and VCRs, but are also available as stand-alone units for professional use.
Telecine   Method for transferring cine film or slides to video.
Tie-clip mic   Very small omnidirectional microphone that is designed to clip inconspicuously on a tie, shirt or jacket.
Tilt   Moving a camera during shooting to point up or down.
Time lapse   Camcorder feature that allows you to take shots that last a few frames at regular time intervals. When played back, the subject movement is greatly increased. Ideal for condensing events that take a few hours down into several seconds of video footage. Also known as an intervalometer.
Timecode   An electronic code recorded on video tape which uniquely labels each frame with a time and frame number so that the video can be edited with utmost precision using a suitable timecode edit controller.
Title generator   Electronic device for creating text and superimposing it on a video picture.
Track shot   Horizontal movement of a camcorder during shooting – such as to move in closer to something, or to follow along with a moving subject.
TTL   Abbreviation for through-the-lens.
Two shot   Any shot in which two people are framed.

Unidirectional   Type of microphone that records sound mainly from in front.

VCR   Video cassette recorder.
VHS (Video Home System)   The most widespread video format in the world, originally developed by JVC and which uses 1/2in-wide tape.
VHS-C (Compact VHS)   VHS tape in a smaller, shorter-lasting cassette.
Video   From the Latin for 'I see', this word must be used with care. Although by definition it refers to the picture alone, it is widely used to mean both audio and video combined (as in video recorder).
Video head   Minute electromagnet used to record audio or video signals on the tape. Several of these are mounted around a rotating drum inside a camcorder or VCR.
Video 8   Sony's trade name for the 8mm camcorder format.
Vision mixer   Device that allows you mix two different signals together so that one gradually fades into the other.
VITC (vertical interval timecode)   Timecode system used on VHS-family camcorders.
VLS (very long shot)   Same as ELS.
VTR (video tape recorder)   Old-fashioned term for a VCR, which stems from the time that recorders used open reels of tape rather than cassettes.
VU meter (volume unit meter)   Used for checking the level of audio signals – found on some camcorders, VCRs and audio mixers.

White balance   System used by a camcorder to measure the colour temperature of a light source and then correct it so that whites, and therefore all other colours, are seen as normal to the human eye.

Wildtrack   Background sound recorded separately from the picture while out on location (often using a separate audio tape recorder) which is later used for audio dubbing at the editing stage.

Wind filter   Electrical circuitry that cuts out much of the hiss and wind noise that might be picked up by a microphone.

Wipe   Special effect where a video picture is gradually covered by a coloured background or another video picture.

Y/C signal   Component video where the luminance (Y) and chrominance (C) are separated. Signal used by high-band recordings.

Zoom lens   A lens with a variable focal length – usually which can be moved from a wideangle setting to a telephoto one.

Zoom ratio   The ratio between the longest and shortest focal lengths on a camcorder's zoom lens.